MG TC
1945-1949

Compiled by
R.M. Clarke

ISBN 0 946 489 009

Distributed by
Brooklands Book Distribution Ltd.
'Holmerise', Seven Hills Road,
Cobham, Surrey, England

BROOKLANDS ROAD & TRACK SERIES

Road & Track on Corvette 1953-1967
Road & Track on Corvette 1968-1982
Road & Track on Ferrari 1968-1974
Road & Track on Ferrari 1975-1981
Road & Track on Fiat Sports Car 1968-1981
Road & Track on Jaguar 1974-1982
Road & Track on Lamborghini 1964-1982
Road & Track on Lotus 1972-1983
Road & Track on Mercedes Sports & GT Cars 1970-1980
Road & Track on Porsche 1972-1975
Road & Track on Porsche 1975-1978
Road & Track on Porsche 1979-1982

BROOKLANDS MUSCLE CARS SERIES

American Motors Muscle Cars 1966-1970
Camaro Muscle Cars 1966-1972
Chevrolet Muscle Cars 1966-1971
Dodge Muscle Cars 1967-1970
Mini Muscle Cars 1961-1979
Plymouth Muscle Cars 1966-1971
Muscle Cars Compared 1966-1971
Muscle Cars Compared Book 2 1965-1971

BROOKLANDS MILITARY VEHICLES SERIES

Allied Military Vehicles Collection No. 1
Jeep Collection No. 1

BROOKLANDS BOOKS SERIES

AC Cobra 1962-1969
Alfa Romeo Spider 1966-1981
Armstrong Siddeley Cars 1945-1960
Austin 7 in the Thirties
Austin Seven Cars 1930-1935
Austin 10 1932-1939
Austin A30 & A35 1951-1962
Austin Healey 100 1952-1959
Austin Healey 3000 1959-1967
Austin Healey 100 & 3000 Collection No. 1
Austin Healey Sprite 1958-1971
Avanti 1962-1983
BMW Six Cylinder Coupés 1969-1975
BMW 1600 Collection No. 1
BMW 2002 Collection No. 1
Buick Cars 1929-1939
Cadillac in the Sixties No. 1
Camaro 1966-1970
Chrysler Cars 1930-1939
Citroen Traction Avant 1934-1957
Citroen 2CV 1949-1982
Corvair 1959-1968
Corvette Cars 1955-1964
Daimler Dart & V-8 250 1959-1969
Datsun 240z & 260z 1970-1977
De Tomaso Collection No. 1
Dodge Cars 1924-1938
Ferrari Cars 1946-1956
Ferrari Cars 1957-1962
Ferrari Cars 1962-1966
Ferrari Cars 1966-1969
Ferrari Cars 1969-1973
Ferrari Cars 1973-1977
Ferrari Cars 1977-1981
Ferrari Collection No. 1
Fiat X1/9 1972-1980
Ford GT40 1964-1978
Ford Mustang 1964-1967
Ford Mustang 1967-1973
Ford RS Escort 1968-1980
Hudson & Railton Cars 1936-1940
Jaguar (& S.S) Cars 1931-1937
Jaguar (& S.S) Cars 1937-1947
Jaguar Cars 1948-1951
Jaguar Cars 1951-1953
Jaguar Cars 1955-1957
Jaguar Cars 1957-1961
Jaguar Cars 1961-1964
Jaguar Cars 1964-1968
Jaguar E-Type 1961-1966
Jaguar E-Type 1966-1971
Jaguar E-Type 1971-1975
Jaguar XKE Collection No. 1
Jaguar XJ6 1968-1972
Jaguar XJ6 1973-1980
Jaguar XJ12 1972-1980
Jaguar XJS 1975-1980
Jensen Cars 1946-1967
Jensen Cars 1967-1979
Jensen Interceptor 1966-1976
Jensen-Healey 1972-1976
Lamborghini Cars 1964-1970
Lamborghini Cars 1970-1975
Lamborghini Countach Collection No. 1

Land Rover 1948-1973
Land Rover 1958-1983
Lotus Cortina 1963-1970
Lotus Elan 1962-1973
Lotus Elan Collection No. 1
Lotus Elan Collection No. 2
Lotus Elite 1957-1964
Lotus Elite & Eclat 1975-1981
Lotus Esprit 1974-1981
Lotus Europa 1966-1975
Lotus Europa Collection No. 1
Lotus Seven 1957-1980
Lotus Seven Collection No. 1
Maserati 1965-1970
Maserati 1970-1975
Mazda RX-7 Collection No. 1
Mercedes Benz Cars 1949-1954
Mercedes Benz Cars 1954-1957
Mercedes Benz Cars 1957-1961
Mercedes Benz Competition Cars 1950-1957
MG Cars in the Thirties
MG Cars 1929-1934
MG Cars 1935-1940
MG Cars 1948-1951
MG TC 1945-1949
MG TD 1949-1953
MG TF 1953-1955
MG CARS 1952-1954
MG Cars 1955-1957
MG Cars 1957-1959
MG Cars 1959-1962
MG Midget 1961-1979
MG MGA 1955-1962
MGA Collection No. 1
MG MGB 1962-1970
MG MGB 1970-1980
MG MGB GT 1965-1980
Mini-Cooper 1961-1971
Morgan Cars 1936-1960
Morgan Cars 1960-1970
Morgan Cars 1969-1979
Morris Minor 1949-1970
Morris Minor Collection No. 1
Nash Metropolitan 1954-1961
Opel GT 1968-1973
Packard Cars 1920-1942
Pantera 1970-1973
Pantera & Mangusta 1969-1974
Pontiac GTO 1964-1970
Pontiac Firebird 1967-1973
Porsche Cars 1952-1956
Porsche Cars 1960-1964
Porsche Cars 1964-1968
Porsche Cars 1968-1972
Porsche Cars in the Sixties
Porsche Cars 1972-1975
Porsche 911 Collection No. 1
Porsche 911 Collection No. 2
Porsche 914 1969-1975
Porsche 924 1975-1981
Porsche 928 Collection No. 1
Porsche Turbo Collection No. 1
Reliant Scimitar 1964-1982
Riley Cars 1945-1950
Riley Cars 1950-1955
Rolls Royce Cars 1930-1935
Rolls Royce Cars 1940-1950
Rolls Royce Silver Cloud 1955-1965
Rolls Royce Silver Shadow 1965-1980
Range Rover 1970-1981
Rover 3 & 3.5 Litre 1958-1973
Rover P4 1949-1959
Rover P4 1955-1964
Rover 2000 + 2200 1963-1977
Saab Sonett Collection No. 1
Singer Sports Cars 1933-1954
Studebaker Cars 1923-1939
Sunbeam Alpine & Tiger 1959-1967
Triumph Spitfire 1962-1980
Triumph Stag 1970-1980
Triumph TR2 & TR3 1952-1960
Triumph TR4 & TR5 1961-1969
Triumph TR6 1969-1976
Triumph TR7 & TR8 1975-1981
Triumph GT6 1966-1974
Triumph Vitesse & Herald 1959-1971
TVR 1960-1980
Volkswagen Cars 1936-1956
VW Beetle 1956-1977
VW Karmann Ghia Collection No. 1
VW Scirocco 1974-1981
Volvo 1800 1960-1973
Volvo 120 Series 1956-1970

CONTENTS

ACKNOWLEDGEMENTS

If you are the type of person who enjoys an introduction to a book, may I suggest you turn to the last three articles and treat them as such. In those few pages you will find a detailed factual article on the marque from Thoroughbred & Classic Cars, Ian Frazer's delightful boyhood reminiscences from Car and finally a beautifully illustrated story from Wheels of an Australian TC by Geoffrey Bewley which unfortunately we cannot do justice to in black and white.

This is the seventeenth title we have produced on MG and the first of a trilogy covering the post-war T series cars.

By the very nature of our books, nothing original lies between the covers. However if you are the right side of 35 much of it will perhaps be new to you and as a consequence interesting.

The aim of Brooklands Books is to make available once again lost information for today's owners, restorers and historians. Our reference series now exceeds 150 titles, none of which could have appeared without the help of the leading motor journals of the world. The enthusiasts who write and publish them understand the needs of hobbyists who cherish their cars and graciously allow us to reprint their copyright articles in small numbers in this way.

Our thanks go to the management of Auto, Autocar, Autosport, Auto Sportsman, Car, Car & Driver, Cars & Car Conversions, Light Car, Motor, Motor Sport, Road & Track, Sports Car World, Thoroughbred & Classic Cars and Wheels for their continued support.

Our thanks also go to Howard Goldman for allowing us to photograph his splendid 1947 TC for the front cover, and also to Moss Motors of Goleta, California who went out of their way to render us every assistance.

R.M. Clarke

A New M.G.
Introducing the TC Midget

M.G. TYPE TC DATA

Present tax..	£13 15s.
Cubic capacity	1,250 c.c.
Cylinders	4
Valve position	Push rod o.h.v.
Bore ..	66.5 mm.
Stroke	90 mm.
Comp. ratio	7.25–7.50 to 1
Max. power	54.4 b.h.p.
at	5,200 r.p.m.
Max. torque	64 lb./ft.
at	2,700 r.p.m.
H.P. : Sq. in. piston area	2.5 b.h.p.
Wt. : Sq. in. piston area	81.5 lb.
Ft./min. piston speed at max. h.p...	3,068
Carburetter	Twin S.U. semi-downdraught
Ignition	12-volt coil, auto advance distributor
Plugs : Make and type ..	Champion L. 10 S., 14 mm.
Fuel pump ..	S.U. electric
Oil filter (by-pass, full flow)	Full flow
Clutch ..	Borg and Beck single-plate, dry
1st gear	17.32 : 1
2nd gear (s)..	10 : 1
3rd gear (s)..	6.93 : 1
Top gear (s)..	5.125 : 1
Reverse	17.32 : 1
Prop. shaft ..	Hardy Spicer
Final drive ..	Spiral bevel
Brakes	Lockheed
Drums ..	9 ins. × 1½ ins.
Friction lining area	104 sq. ins.
Car wt. per sq. in.	16.53 lb.
Suspension ..	Semi-elliptic Luvax Girling dampers
Steering gear	Bishop cam
Steering wheel	Bluemel adjustable. 17 ins.
Wheelbase ..	7 ft. 10 ins.
Track, front	3 ft. 9 ins.
Track, rear..	3 ft. 9 ins.
Overall length	11 ft. 7½ ins
Overall width	4 ft. 8 ins.
Overall height	4 ft. 5 ins.
Ground clearance	6 ins.
Turning circle	37 ft.
Weight—dry	15.5 cwts.
Tyre size ..	4.50 × 19
Wheel type	Wire 19 × 250 W.B.
Fuel capacity	13 gallons
Oil capacity	10½ pints
Water capacity	14 pints
Electrical system ..	Lucas 12-volt (earth return, automatic voltage control battery charging)
Battery capacity ..	51 amp. hr. capacity 10 hr. rate

Top Gear Facts:

Engine speed per 10 m.p.h.	632 r.p.m.
Piston speed per 10 m.p.h.	374 ft./min.
Road speed at 2,500 ft./min. (piston) ..	67 m.p.h.
Litres per ton-mile ..	3,050

Weights dry. (s) synchromesh.

THE immediate post-war programme of the M.G. Car Co. is to concentrate on one model only, to be called the TC Midget. This car is a direct derivation from the TB model, but, as the latter car was introduced only shortly before the war and made in but limited numbers, it is our purpose to give a detailed description. In this we hope it will be borne in mind that many of the remarks relating to the new TC type apply also to those TB models which are already on the road.

Birthday Presentation

This year is the 21st birthday of the M.G. Co., and during these years cars of this make have won 44 major races, have broken International class records on 121 occasions (including the first cars of their size to exceed 100 m.p.h., two miles a minute, three miles a minute and 200 m.p.h.), and almost innumerable firsts and places in reliability trials and club events. The overwhelming majority of these wins were scored by the earlier series of four- and six-cylinder cars with overhead camshaft engines. In view of their successes it was something of a shock when this type was abandoned in 1937 in favour of a push-rod unit of greater capacity and lower specific output, but, although the push-rod valve system is used, the TC is a distinct reversion to the earlier genre. Peak power is reached at over 5,000 r.p.m., and with a compression ratio of 7½ to 1 the very creditable output figures of 2½ horse-power per sq. in. of piston area and nearly 45 b.h.p. per litre are realized. With approximately 55 b.h.p. available at the flywheel, the road speed of the car can be taken as 80 m.p.h.

plus or minus, say, 2.5 per cent., according to position of screen and direction of normal wind. At 5,500 r.p.m. the speed on the indirect gears is 64 m.p.h., 45 m.p.h., and 26 m.p.h., respectively. In conjunction with 70 horse-power per ton on the bare weight of the car, the all-round road performance is manifestly very much higher than average.

Turning now to the car as a whole, only one body-type is at present available, this being a two-seater from what may be termed "typically Midget" style. Although similar in outline to TA and TB models, it is, in fact, 4 ins. wider over the seat than the last named, and this makes a very considerable difference to passenger comfort, particularly when the hood is raised and sidescreens are in position.

Increased Comfort

The latter components have received particular attention. They are given very stout frames and fit really closely to the hood and to the windscreen pillars, as can be seen from a photograph. When fully closed the car has a smart coupé-like appearance, and the comfort when driving does not belie the external aspect. Signalling flaps are fitted to the front and side screens (there being no direction indicators), and it is thought that even in really cold weather it would not be necessary to wear an overcoat. In the extreme rear of the body is a locker with a hinged lid to contain the side screens when they are not used. This should prevent undue and premature scratching of the celluloid panels, and if good care is taken the car should be capable of giving excellent service both as an open sports two-seater and as a closed model.

Introducing the TC M.G. Midget—Contd.

In accordance with usual M.G. practice, the 13-gallon fuel tank is placed externally to the body, and has a quick-release filler cap. The capacity gives a cruising range of over 300 miles without calling on the reserve supply. On the TC model there is no reserve tap, but a light indicator on the right hand of the facia panel. This also contains large-diameter Jaeger speedometer and tachometer, in addition to the other conventional instruments. There is now a single position for dynamo charging, as the output from this component is controlled by a constant-voltage unit. The accumulators, which on previous T series cars were placed somewhat inaccessibly

order to provide the excellent wheel lock which is a valuable feature of a trials car.

The springs as a whole are provided with limited motion, and the practical results on the road are a faithful interpretation of the theoretical anticipa-

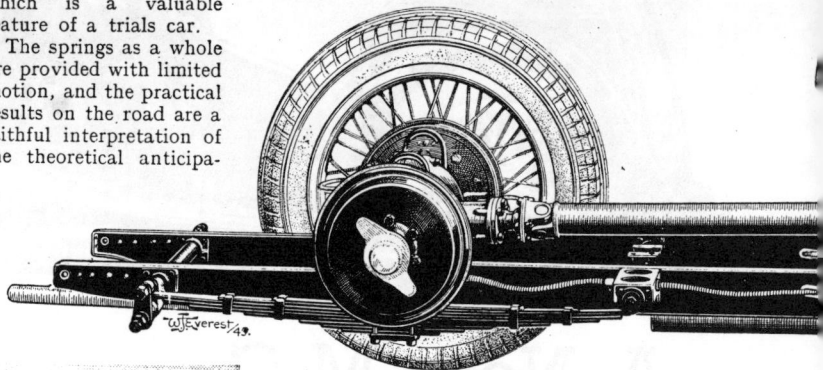

THE WORKS.—This drawing, copyright by "The Motor, by the large pressed steel dash. On the TC model diameter Bluemel spring wheel with adjustment for col voltage control; the oil pump, which

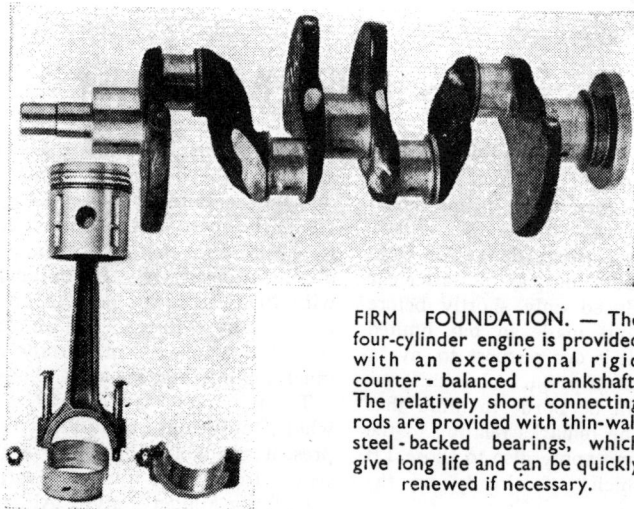

FIRM FOUNDATION. — The four-cylinder engine is provided with an exceptional rigid counter - balanced crankshaft. The relatively short connecting rods are provided with thin-wall steel - backed bearings, which give long life and can be quickly renewed if necessary.

to realize that this power unit has been specially developed for use in this particular car, and embraces a number of meritorious features. In particular, bearing design and stiffness in all the rotating and reciprocating parts has been aimed at. The stroke/bore ratio of 1.35 to 1 is distinctly below average for British automobiles, and the crank itself is very robust and well counter-balanced, as can be seen from an illustration. The connecting rods are short (the centre lengths being under four times the crank radius), and for both main and big-end bearings the thin wall white-metal type of unit is employed.

below the floorboards of the rear luggage compartment (and thus subject to neglect), are now placed under the bonnet in the pressed-steel dashboard assembly. The rear of the body is reserved as a luggage compartment, and the two front seats have a single squab, the rake of which can be varied. There is also a range of fore and aft adjustment.

The steering column is the Bluemel extensile type, which can be moved by slackening a locking bolt. The screen is, as one would expect, of the fold-forward type, the windscreen wiper motor being placed at the top and driving two arms.

Chassis Features

The body is mounted on a channel-section chassis, which has a number of interesting features. In particular, the frame is carried beneath the rear axle with the flat semi-elliptic springs outrigged and also passing beneath the axle. They are supported at the rear on a tubular cross-member which passes from side to side to act as a frame-bracing member. The shackles are now of conventional type fore and aft, and the springs front and back are controlled by the latest type of Luvax-Girling damper. The rear springs are out-rigged to the greatest possible extent to provide resistance to roll, and, on the other hand, the front springs are relatively closely placed in

tions with this type of layout.

The rear axle has the conventional spiral-bevel drive and is connected with the gearbox by a Hardy-Spicer propeller shaft. Following racing experience, this is of particularly large diameter, so that there can be no question of it running at a critical whirling speed within the limits of the road speed of the car.

Synchromesh on Three Gears

The gearbox has a remote-control gear lever and provides synchromesh on all the upper three ratios. This is, again, a comparatively new arrangement for the T series, and one which makes possible very slick gear changing without double clutching. Going down, the procedure is to hold the throttle open and steadily press the lever towards the desired lower gear position, whilst changing up can be done with great rapidity. Against this, quiet changes without the clutch require a much higher than average degree of skill if they are to be brought off successfully. Power to the box is transmitted by a dry single-plate clutch of Borg and Beck manufacture; the flywheel is much lighter than on the TA engine, with a correspondingly more rapid response to throttle movements.

Reference has previously been made to the r.p.m. and output characteristics of the engine. It is important

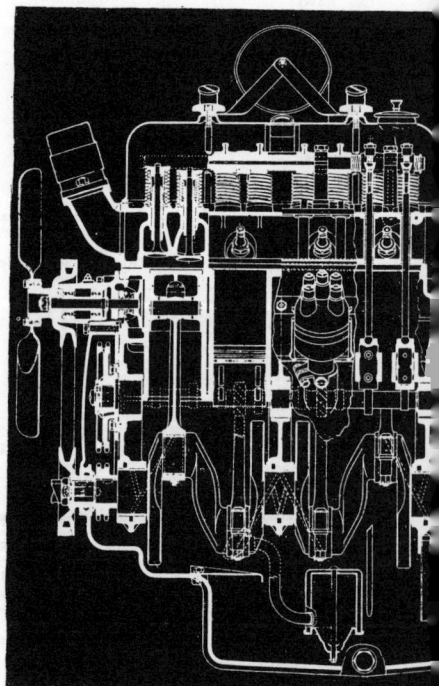

INSIDE INFORMATION.—These drawings develops nearly 55 b.h.p., representing th b.h.p. per s

the engine is set well back in the frame, the latter being stiffened amidships
es toolbags and accumulators. The well-raked steering column carries a large
Right) The near side of the engine. The large capacity dynamo has constant-
rough a floating pick-up, delivers to an external full-flow filter.

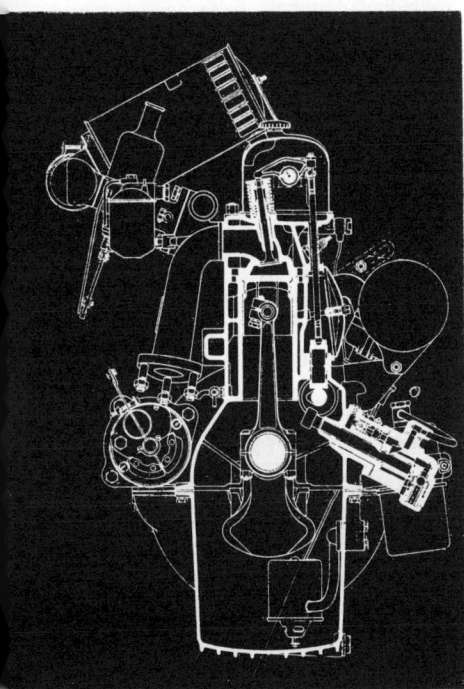

eral arrangement of the TC engine, which
table output of 45 b.h.p. per litre and 2.5
piston area.

A large-capacity oil pump feeds the lubricant through a full-flow filter. A relief valve limits the oil pressure between 40 and 50 lb. per sq. in., but the capacity of the pump is such that this figure can be well maintained throughout the useful life of the bearings. Clean oil is also ensured by the use of a floating oil pick-up in the deep, ribbed sump, which contains over a gallon of oil. The camshaft is driven by a duplex chain and the push-rods are shortened by placing the lower ball and cup at the top of the tappet. Adjustment is provided on the rocker and the slightly inclined valves are controlled by double-valve springs.

As can be seen from the cross-sectional engine drawing, there are ample water passages around the valve seating and the cooling water is specially directed around the cylinder head. The water from the pump is carried through a conduit cast into the manifold side of the block, and from this is taken into the rear of the cylinder head. The flow of the pump is, therefore, directed almost entirely through the head, but there are passages drilled between the head and the block so that the latter receives virtually static water. This has the effect not only of keeping the head temperature down, but keeping the block temperature up, thus reducing wear on the cylinder bore and maintaining high mechanical efficiency.

Even Cooling

Additional to this feature, the normal thermostat, cutting out the circulation of water to the radiator during the initial warming-up period, is retained. A fan is also employed in the cooling system, the water pump being mounted on the fan-drive shaft on the front of the cylinder block. As can be seen from the sectional drawing there is a water space between each cylinder bore, a particularly desirable feature on a high-output engine.

Light-alloy Aerolite pistons are employed, there being two compression and one slotted scraper ring. The inlet valves are approximately 10 per cent. larger than the exhaust valves and receive mixture from a pair of semi-downdraught S.U. carburetters. These have a manual mixture control and long pipes leading from the flow-chamber vent, so that if for any cause flooding should occur there would be no danger of fire as a result of fuel dribbling over the exhaust manifold.

The latter has four branches arranged in Y formation to join a single tail pipe. Each carburetter is joined to a light-alloy manifold connecting to a single air silencer and cleaner.

On the exhaust side a single Burgess silencer is used with a flexible connection in the exhaust pipe to take care of engine movement on the rubber mountings.

Ignition is by 14 mm. Champion plugs, which are inserted at an angle in the cylinder head on the side opposite to the manifolds, a feature which obviously assists in plug maintenance, and irrespective of type and position. This is in turn a matter of considerably more importance on high-output engines than on less developed and more soberly driven types. The advance and retard mechanism is governor-controlled so as to vary with engine speed, the distributor being driven by a skew gear off the camshaft.

High Mechanical Efficiency

The comparatively high specific output has not been obtained at the expense of low end torque. On the contrary, the power curve reveals a b.m.e.p. in excess of 100 lb. at 1,000 r.p.m. and over 120 lb. between 1,700 r.p.m. and 4,200 r.p.m. These figures have been secured not only by careful porting and camshaft design, but also by maintaining a better than usual mechanical efficiency over a wide speed range. The figure on this count reaches 90 per cent. at 800 r.p.m., and is held to 80 per cent. at 4,000 r.p.m., which is the equivalent of 2,350 ft. per minute piston speed.

As one might expect, these factors reflect an excellent specific consumption on full throttle. This is less than 0.52 pint per b.h.p. hour over a speed range 1,500-4,500 r.p.m., and only rises slightly to 0.55 pint per b.h.p. hour at the peak speed of 5,400 r.p.m.

As can be seen from the data panel, the gearing is such that 67 m.p.h. can

Introducing the TC M.G. Midget—Contd.

be maintained at 2,500 ft. per minute piston speed, and at this velocity the engine will be working at approximately 60 per cent. of full throttle. It is, therefore, reasonable to suppose that this car is fully capable of a sustained cruising speed of 65-70 m.p.h., where road conditions permit, without overstressing the engine.

ALL-WEATHER.—Fully open or adapted for winter wear.

From the factors of horse-power per ton and litres per ton mile, as previously mentioned, one can expect a 0-60 acceleration time of approximately 22 seconds, and 10-30 m.p.h. acceleration time of approximately 10 seconds. Although it is at present difficult to obtain accurate figures, tests made after calibrating a production-type speedometer indicate that the former figure can be reached, and that the latter is 12.2 seconds with a carburetter set on the lean side and using existing Pool petrol. The acceleration time in top gear between 30-50 m.p.h. improved substantially to 9 seconds.

Hydraulic Brakes

The high performance provided by this power unit is matched by the stopping powers of the Lockheed brakes, which have cast-iron drums of 9-in. diameter with deep stiffening ribs. Due to the comparatively low all-up weight of the car, the loading per sq. in. of Ferodo brake-lining area is about 25 per cent. less than usual

VARIABLE REACH. — The range of adjustment of the Bluemel steering wheel.

practice for conventional saloon cars, and the brakes can be used heavily without giving trouble from fade. Directional control is by Bishop Cam-gear working a transverse track rod and designed to provide reasonable high gearing with lightness of control at low speeds.

Control Gear

The handbrake is of the racing "fly-off" type; that is to say the ratchet is engaged by pressing down the knob on the top of the handle, release being obtained by pulling back and letting go. This brake is connected to the rear shoes by enclosed cables and provides a really powerful stopping effect. Thus, it is not only useful in emergency but also materially assists handling under competitive conditions. The clutch and brake pedals project vertically through the floorboards with dust excluders, an arrangement which gives very comfortable pedal action when employed in conjunction with comparatively low seating position. Moreover, the dash assembly can be brought down to the frame without slots being needed for the pedals, a feature which materially assists in excluding draughts, noise and engine fumes.

The wheels are Dunlop knock-off R.W. wire type, and are worthy of note in that they are considerably larger in diameter than the present fashion. The technical arguments in the large versus small wheel controversy are evenly distributed, but there can be no doubt that the former score very much in the aspect of appearance. In this case they undoubtedly contribute towards the well-balanced aspect of a high-performance small car, and one that will doubtless give great pleasure to a large number of sporting motorists.

EXPERIENCE TEACHES. —As can be seen from this drawing, showing the position of the principal chassis components in relation to pedals, seating and steering, the design of the TC Midget owes much to competition and racing experience. In these endeavours the Company have had exceptional successes during the past 21 years.

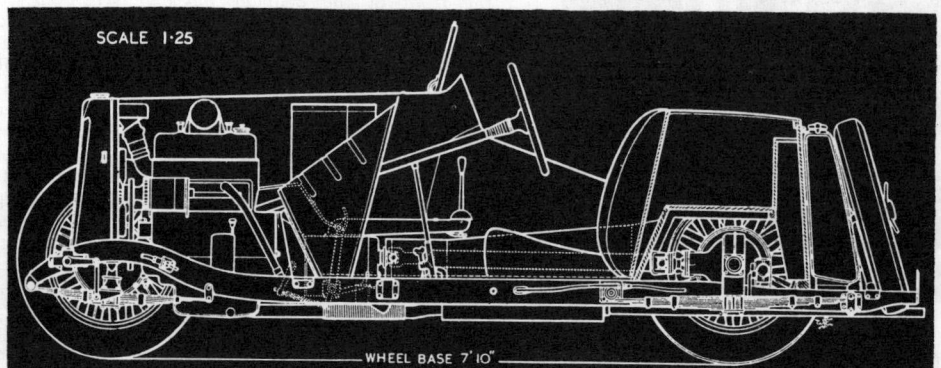

SCALE 1·25

WHEEL BASE 7' 10"

Safety Fast !

THE MG CAR COMPANY LTD. ABINGDON-ON-THAMES, BERKS

"Hill Busting"

The Behaviour of a Post-war Sample

Specification

Engine.—10.97 h.p., four cylinders, 66.5×90 mm. (1,250 c.c.). Push-rod operated overhead valves, three-bearing counterbalanced crankshaft, aluminium alloy pistons, steel connecting rods, twin S.U. carburettors, pump water circulation with thermostat, forced oil feed with pressure filter.

Transmission.—Single-plate clutch, four-speed remote control gear box with synchromesh on top, third, and second. Ratios: Top 5.12, third 6.92, second 10.1 and first 17.32 to 1. Hardy-Spicer balanced propeller-shaft to spiral bevel three-quarter floating rear axle.

Steering.—High-geared cam type.

Suspension.—Half-elliptic springs; rubber bush bearings, Luvax-Girling hydraulic dampers.

Brakes.—Lockheed hydraulic; cable hand brake and quick-release lever.

Tyres.—Dunlop 19×4.50in. centre-lock wire wheels.

Fuel Tank.—13½ gallons.

Dimensions.—Wheelbase 7ft. 10in., track 3ft. 9in. Overall length 11ft. 7½in., width 4ft. 8in., height 4ft. 5in.

Price.—£375 plus £104 18s. 4d. Purchase Tax.

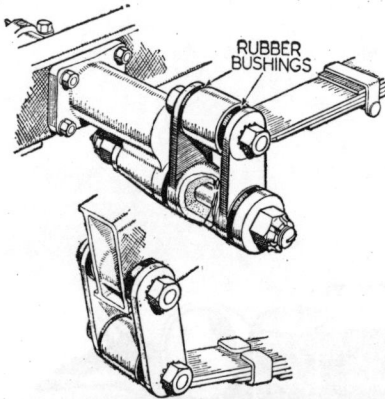

RUBBER BUSHINGS

New shackles have been adopted and the springs are fitted by means of rubber bushes.

THIS is an occasion calling for something a little different from the orthodox description of a car returning into production. The reason is that the M.G. Midget occupies a unique niche in the affections of skilful members of the motoring public. This affection has its origin in two sources. First a proper appreciation of the high-speed histories written by 100 per cent. British M.G.s on the scroll of world's and class records; secondly, the natural pent-up longing of enthusiasts, particularly those in the Services, to experience again, or maybe for the first time, the thrills of safe, fast motoring. They are to feel the joy of holding a fast little car to a true course; to time the gear change exactly to the engine revs.; to feel the breeze in the hair; to have the human pleasure of being envied; to hear the exhaust note rising; in short, to experience the joys of the open road, and to improve skill by exercising it. These things bring closer the old excitements of the speed trial, the reliability trial, the rally, and all competitions.

The "Midge" offers so much charm at, purchase tax apart, a moderate price. It is so well thought out for its purpose;

and, I might add in wistful requiem, so entirely characteristic of the great little man who originated the M.G.s.

I have just enjoyed what so many readers of this journal would regard as a notable stroke of luck, the use for a day on the road of a new post-war M.G., the Midget Type TC. Before going further I had better explain that the TC has practically the same mechanical specification as the TB, but the body is considerably wider and the general comfort much improved. Details of the changes are tabulated.

Golden Opportunity

Inspiration on the best use to be made of a golden opportunity came from the recollection of recent correspondence with a reader who wanted a lot of details about the various hills in the Cotswolds which in years gone by figured in the earlier forms of *The Autocar* routine of Road Tests, and which at one time were included in the Colmore Cup Trial. Fifteen or twenty years ago I used to know those hills like the back of my hand, as I had to climb them about once a week, on cars that sometimes made fairly heavy going of the job. I felt that it might be interesting to hunt out those hills again and see what the "Midge" thought about them. We were lucky in having ideal sports car weather—warm and no rain.

One sits low in the Midget; in fact, a hand on a longish arm can touch the

The new Luvax-Girling hydraulic dampers have further improved the suspension.

ground, and the steering wheel is close to the vertical plane. The driver can tuck into the job. I found it quite easy to adjust the seating position to my particular needs. The cushion, wider than of old, can be adjusted fore and aft for leg reach, and thereafter the squab can be adjusted separately by quadrant and locking pin to give the required angle. One should sit fairly upright in a car of this type; it gives better control and visibility, and is less tiring on long fast runs. The spring-spoked steering wheel is telescopically mounted and is therefore instantly adjustable for reach.

After six years of very restricted motoring in a softly sprung comfortable saloon, sticking to a painfully slow 30 to 40 m.p.h. with one eye always on the

With hood and side screens ap

petrol gauge, taking over a Midget again is quite an exciting experience. The seating position feels a bit strange. I later noticed something else. When I got back into my own familiar car the day after, I felt stranger still, as if I were high in the air, and not sure of my control of the steering, which does rather confirm that the driving position beloved of sports car drivers certainly has more to it than mere appearances.

However, to return to the Midget. Left of centre of the instrument panel is a pull-out starter plunger, and next to it a mixture-enriching plunger. After switching on with the ignition key you pull the plungers outwards with each hand, and let go as soon as the engine starts, as it does easily enough from cold. Incidentally, the horn button and dip-switch are mounted on the instrument panel close to the left side of the steering wheel. The hand brake is of the racing type, meaning that when starting from rest you pull the lever towards you and let go and the brake is released. If you want the ratchet to grip you press downwards on the knob at the top of the lever while you are pulling the brake on, and the lever stays where you leave it.

The short gear lever is mounted on a tunnel which positions it well back in the car, conveniently close to the left hand.

n a 1945 Midget

.G. Midget Over the Cotswold Hills

By Montague Tombs

her-proofing is achieved and smart
ntained.

Fine-weather trim of the new M.G. Useful luggage space is available behind the seat.

at first, and it was not for some time that I realised that the engine has a good torque at low revs, and that the car can be handled as sedately on top gear as any town carriage. In short the temptation is to let the Midget rip; but you don't have to.

At first the steering feels a trifle quick, but after a few miles one realises that it is just right for the car, for not only does it allow for a quick manœuvre, but the back end of the car follows the front end exactly, so that you can place it on a curve and hold it there, with no two bites at the steering cherry. The brakes also are good; one need not be chary of using them at high speed, and the pedal pressure is not heavy. I am not going to say what the maximum speed of the TC is, because I frankly do not know, having had no private measured quarter-mile for test. But I have it on the

authority of the M.G. test and experimental department that this car with the screen up has been timed to cover the quarter-mile at 78.94 m.p.h., with a mean speed of 78.26 m.p.h. I am quite prepared to believe it, not only because I have handled the car, but also because I can realise the awful temptation there must have been to call it a round 80 m.p.h.

Screen Position

Mention of the windscreen reminds me of something. The screen can be folded flat forwards when desired, and when up it reaches a stop which causes it to register for the fastening of the folding quick-lift hood. I would like to offer the suggestion that when the screen is up it seems a shade too near the vertical to sweep the air flow clean over the heads of the occupants of the seats, and a back draught is noticeable. If it were made possible to incline the screen a trifle further backwards when the hood is not in use I fancy that the back draught could be avoided. Also the present angle does entail a certain amount of "ghost reflection," for a person of my small stature, anyway.

In the matter of cruising speed the car settles down to a very comfortable 55 to 60 m.p.h., which it seems to enjoy as much as the driver. It also likes on occasion to potter at 35 m.p.h., when it takes the ordinary sort of main road hill in its stride on top gear. You will probably want to be told what the gear

ratios are. The answer is top 5.12, third 6.92, second 10, and first 17.32 to 1. The engine peaks at 5,250 r.p.m. On top gear the road speed is 15.84 m.p.h. per 1,000 r.p.m. The tyres are 19×4.50in. Dunlop on centre-lock wire wheels. I had a puncture from a wire nail, so I know that the wheels are centre-lock. It is nice to meet them again—they are not half so much bother to change as "bolt-ons."

Meanwhile we had been getting to those more or less deserted hills which the title of this article says we were going to bust. They are all in the terrain surrounding Broadway and Winchcombe. The first before Broadway was Saintbury. In the first place I had forgotten the details of the hill; in the second I had not learned how to handle the latent possibilities of the Midget in speed and cornering. So I changed down at the

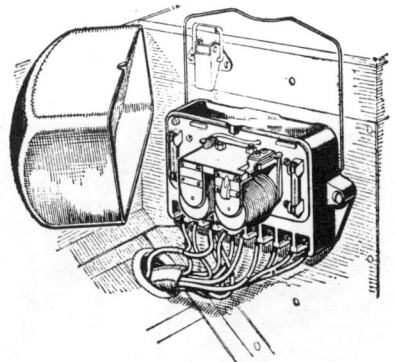

Improvements and Modifications Summarised

Two-seater body made approximately four inches wider; seats increased in size, adding to comfort and appearance.

Suspension: rubber bushes fitted to springs, and new Luvax-Girling dampers adopted. Noticeable improvement in riding without loss of stability.

Battery: single 12-volt, accessibly mounted in dash bulkhead under the scuttle.

Tool locker of large size mounted behind battery box; accessible from each side.

Instruments modified and brought up to date. Electrical fuel gauge shows green warning light when fuel reserve is in use.

"Under-bonnet" appearance improved. Electrical leads and pipe lines neatly grouped. Engine finish in grey cellulose instead of black.

Paintwork in preliminary production will be black with choice of trim in red, green, or biscuit.

Electrical equipment: improved and simplified two-fuse wiring.

Features of coachwork: Fold-flat screen, large luggage container behind seats. Felt-lined compartment for side screens. Quick-lift hood folds flat inside body, and is enclosed by a tonneau cover.

As there is synchromesh on top, third, and second, gear changing is completely easy, provided that you push the clutch right out and move the gear lever progressively—that is, without snatching at it.

Well, you have the engine started, the brake off, bottom gear in, and off you go. You change up into second, and then third, but if you happen to be starting off in a town you may not be quite sure of yourself at that moment, i.e., ought you to stay on third and keep up the engine revs. ready for acceleration in traffic, or change up and behave like a saloon driver? The reason for this first instinctive debate is that the car feels mettlesome, ready to be off in a flash, and inclined, as it were, to dance on its toes. That is because the suspension at low speeds has a quicker return than the slow motion to which one is more accustomed, and because the engine is very quick to respond to the accelerator pedal.

Perhaps it is natural to look for and find first, in the Midget, a dashing sort of performance, smartness off the mark, and quick acceleration by revving up a bit on the gears. The car took me that way

Near-side view of the engine, which is readily accessible. A light finish adds to the smartness of the power unit.

The neat fuse box. Wiring has been simplified to a two-fuse system.

wrong places with faultless misjudgment, and had to save face by calling the attention of my Better Half to the scenery.

According to the mileage recorded the section of Saintbury I covered was exactly one mile long. The gradient averages about 1 in 11.9, but the steep stretch in the middle is 1 in 6.1. The approach is past a few houses, leading to a left sweep, then round a fairly sharp right-hand bend on to the 1 in 6.1: surface indifferent. I took the approach at about 50 on top; it seemed so easy that I hung on to top round the right bend, changed late and climbed the steep on third at about 32 m.p.h. What I should have done was to change at about 45 before the bend, and roar up the steep. Sorry! These hills seem to have become less steep with the passage of time, or else the Midget doesn't think much of them. The way that car will hang on to 30-35 on top gear rather tempts one to err.

Next on the list was Willersey, which joins Saintbury near the top. We covered point nine of a mile from start to join. It is a tricky hill. You start off easily on a gently increasing grade and keep a good speed, then round a wide right-hand sweep and suddenly reach a sharp ingrowing bend. A nasty one. I changed on the sweep to third at 45, got round the sharp bend at 35, not fast enough, and had to drop to second to climb the stretch of 1 in 5.9 at 32 m.p.h. By this time I was getting to know the car a bit better.

Everybody knows Fish Hill; it leads out of Broadway to Moreton-in-the-Marsh. It has an average grade of 1 in 11, a good surface, and measures about 1.7 miles long. It has some easy sweeps, one lovely left bend about half-way up, and a right bend near the top. The scenery is marvellous. Large lordly cars with puissant power-to-weight ratios sometimes climb Fish in stately fashion on top gear. Ordinary cars scrape up on third if the driver is good on the bends. We climbed to the first bend at 50 or so on top, changed for the bend, swung round it at 34, rose to 45 for the rest of

the journey up to the top bend, and rounded that at about 36.

Of course when you get to the top of Fish you have to explore the pretty bit to your right, and show your Better Half that gem of a village Snowshill. That led me close, too close, to once favourite bits of colonial going on the flanks of the hills above Stanton and Stanway. So I went down Stanway main road hill, which is quite beautiful, with that unmatched charm of the Cotswold country, and I fell for the temptation of once again climbing Old Stanway.

Not Advisable

Perhaps you don't know Old Stanway. If so, be advised and don't explore it on wheels just now. It is a sort of forest track leading up between some cottages at the foot of the normal Stanway Hill. Because it is cart-rutted and cross-gullied they used to include it in trials. I always loved it, for it is a charming place where you could picnic on a Bank Holiday and never see a soul. I don't love it any more. I set the Midget to it without walking up first. I blotted my copy book. The tracks have become so deep and the cross gullies so abysmal that, despite trying to pick a dainty course, I bumped the

The hood folds flat and well below the tonneau cover. A felt-lined compartment houses the side screens.

Controls are neat and symmetrically arranged. The rev counter and speedometer have large, easily read dials.

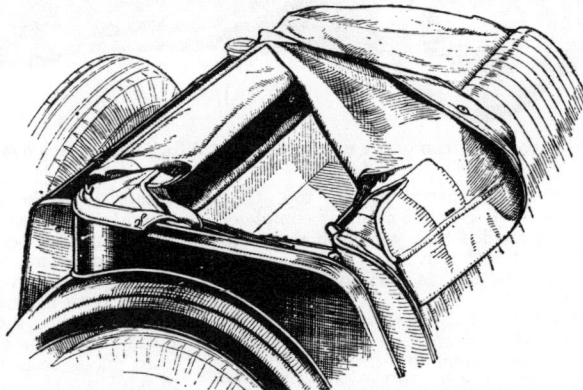

silencer on a rock and split it . . .

When you journey from Broadway to Cheltenham you pass through Winchcombe, where there is a fine old church supporting quaint gargoyles wearing top hats. Why top hats? There is also a turning to the left in the middle of the main street, which leads to Sudeley Hill. This is long, about 1.2 miles, starts easily and grows steeper and steeper, with no bends, and near the top there is a longish stretch of about 1 in 5. We climbed the first part at 49 on top, dropping to third at 43, and then up the steep at 26 on second.

From Winchcombe we ran on towards Cheltenham, for I had a fancy to see Gambles Lane, or Rising Sun as it used to be called. That was once a defeatist hill, said to be 1 in 3½ on the short steep at the top. Many a car in early competitions konked out on Gambles Lane, but the surface was poor then, whilst now it is quite good. The Midget raced up that steep bit on second gear at 26 m.p.h.

On the way home I thought of trying Edge Hill, once beloved of Coventry car testers. Eventually I reached Edge, on the Kineton to Banbury road. It is about three-quarters of a mile long and has a maximum grade of 1 in 7 from about half way to the top. There is a

left curve at the foot, a steeper pitch, and then a bend to the right, whence the grade grows to its maximum. It is not a hill that can be rushed. I changed to third below the bend, rounded it at 43 and climbed the rest at 45 to 43 on third. An easy climb.

Now to sum up my impressions and conclusion. The conclusion is that if you want to go hill busting with an M.G. it is desirable to choose something stiffer than this chicken feed, which cannot bring it below second gear. The impressions are that the "Midge" quite justifies the old motto of "Safety Fast." The little car is rock steady and stable, and handles with a satisfying accuracy. It is quite charming to drive fast. But it also has another merit: you can drive quietly and peaceably when you feel like it, and do everything on top gear. The engine is smooth and most willing. It will rev up to the limit or pull sweetly at low speed. And all the time it is well up to its job. It never suggests that it is being over-driven. In short the car is an extremely well-balanced design.

HOMEWARD TREK—and a glimpse of the big rear tank and the convenient mounting of the spare wheel.

One of the most popular of British sports cars is rolling off the assembly line again—

The M.G. Midget

FOR dyed-in-the-wool sportsmen, or folk who prefer rapid motoring even if they have not been bitten with the trials fever, one of the most encouraging releases of the past month was the M.G. Midget, and we lost no time in making a special journey to Abingdon to see it and to experience the exhilaration of driving it on the open road. Later on, we shall test it over a more extended mileage and report on its performance in greater detail. For the present, we are content to say that, after a black-out of six years, the M.G. Midget has emerged outwardly little changed, inwardly refreshed.

The outward change centres principally in the body, now several inches wider, so as to permit of greater elbow room—which, in its turn, means an even better sense of control. The spring fixings, too, have received attention, and we noted with interest that the new Luvax-Girling damper is fitted all round. The effect of all this is noticeably better road holding, particularly on corners.

The new car (Series TC) is a real thoroughbred with lashings of power under the control of the driver's right foot; yet, if he wishes it, as docile as a town carriage. To drive the M.G. Midget, even for only 20 miles or so, is to recapture the real spirit of the open road, and although it may rain—as it did during our run—one shirks not the thought of getting wet, but of putting up the hood. There again, however, the M.G. scores, for if the weather beats you in the end, the hood and side screens make things thoroughly snug and waterproof.

As always, the safety of the M.G. is an obvious quality. There is a feeling of perfect security even when one's foot is hard down on a winding unknown road. Stability, good steering and powerful Lockheed brakes are contributory factors.

Before we took our departure from Abingdon we were conducted to one of the main assembly bays, and our eyes were gladdened by the sight of some 40 or 50 Midgets, half of them on the " down " line; that is to say, in chassis form, and the other half on the " up " line, being fully clothed with body, bonnet and all the rest. In this particular shop the transformation from war to peace is nearly complete. Difficulties on the supply side are being overcome, and in a few months there should be a steady flow of cars off the assembly lines which will do something towards satisfying the hordes of enthusiasts whose first love was an M.G. and whose affections will never waver.

The specification of the TC model presents many little points of interest to the sports car connoisseur. For example, two S.U. semi-downdraught carburetters cater for the four-cylinder, 1,250 c.c. engine; the sump holds 1½ gallons of oil and is ribbed for cooling; engine temperature is controlled by an automatic thermostat. Then there is synchromesh on top, third and second, the top gear ratio, incidentally, being 5.1—delightful for fast cruising. The third gear ratio is 6.9, which means that there is no delay on a steepish gradient, even if you have to notch down a peg.

If a long, non-stop run is the order of the day, the designer has thoughtfully provided a 13½-gallon petrol tank, whilst, to make certain that your position is just right for that kind of motoring, the seats have sliding adjusters, the squab is adjustable for angle and the steering column is telescopic, so that, within limits, you can arrange the wheel just where you like it. The M.G. invites long-distance travel for two, but that in turn involves adequate luggage space. The Midget has it—just behind the seat squab.

Finally, one thing will strike the student of car design. The seat is nearer the back than the front axle, which surely stands for a very nice distribution of weight and presents a style of body which is so essentially part and parcel of the make-up of a real sports car.

IN BRIEF

Engine: Four-cylinder, o.h.v., 66.5 mm. by 90 mm., 1,250 c.c., 10.97 h.p., pump cooling, two S.U. carburetters.

Transmission: Single-plate clutch, four speeds, 5.12, 6.92, 10.0 (with synchromesh) and 17.32 to 1., Hardy-Spicer balanced prop. shaft.

General: Lockheed brakes, S.U. petrol pump, 12-v. electrical equipment, tyres 19 ins. by 4.5 ins., wheelbase 7 ft. 10 ins., track 3 ft. 9 ins.

OUTWARD BOUND in Sutton Courtney. This three-quarter front view conveys a good impression of the wider body of the new M.G. Midget.

CONTROL ROOM. The rev. counter, not the speedometer, is under the driver's eye. Note the stubby gear lever and the racing-type hand brake

MOUNTAINS IN THE MIST.—A view from the top of Trinafour looking towards the roads to Tummell Bridge and Kinloch, Rannoch. This road is rather rough for the few miles from Dalnacardoch to Trinafour, but improves from there on and is well worth trying.

The Last Week-end in June

IT all started with a telephone call from the M.G. car Co.

The new "TC" Midget which was announced last October is now in steady production, and it occurred to the executives of the Company that some hard driving of the current model by a member of the technical Press might give them the private owner's point of view. In pre-war days such test runs were pretty general, but what with the petrol shortage and the lack of cars very few organizations have been able to plan much of this sort of research since the war ended.

It so happened that a typical example of the TC Midget could be spared for a long week-end, and it was felt that something in the nature of 1,000 miles of fairly constant driving would supply the factory with some valuable data. From the point of view of "The Motor" this also seemed to be a wonderful chance to survey the Great North Road and the Highlands once again, so that on the afternoon of Thursday, June 27, preparations were made for the sort of journey that any motorist would regard as a heaven-sent opportunity.

An editor finds it difficult to avoid being chairborne. In common with almost everyone else, neither my wife nor I have covered any great mileage in a good civilian car for several years. Our interest in the whole event was heightened by the fact that my wife had a trouble-free run with a P-type Midget in George Eyston's team at Le Mans in 1935. Consequently, her reactions to 10 years of development could be considered interesting.

We set out from London on the Thursday at 5.30 p.m. and motored steadily north for a distance of 63 miles, when a halt was made at "The George" at Buckden for dinner. This preliminary drive was valuable from the point of view of getting to know the car. Unfortunately, the weather deteriorated rapidly and both hood and side screens were brought into use in due course. In connection with this important matter of weather protection, quite a lot should be said about the new Midget. As always one sits very much in and not on the car, and as the body width has been increased by 4 ins. in the post-war model, the driver's right arm can be used either with the elbow over the door or brought inside with equal facility. The mudguarding is first class in its efficiency, so that the Midget can be driven through pretty bad weather with the hood down. However, with the prospect of an all-night run and another two-days' motoring after that, plus occasional stops for food, discretion seemed the better part of valour, and it was thus that we discovered that the new hood and side screens do their job in a very satisfactory way.

It would be difficult to think of a better form of small-

London to Inverness and Back in a "TC" Midget

By Christopher Jennings

car transport for two people on this sort of journey than the M.G. The luggage container held two suitcases and an assortment of coats, camera and picnic basket. Even so, the elastic-tensioned waterproof cover fitted neatly into place and as the side screens have a separate locker they can be stowed without the risk of finding them damaged beyond repair by the luggage at the end of a fast run over bad roads.

Performance of the car itself showed great promise. The four-speed gearbox produces about the fastest normal change in existence. A maximum of 64 and 45 m.p.h. on third and second ratios, respectively, plus the 1,250 c.c. engine delivering 55 b.h.p., means that the Midget can produce, when required, a degree of acceleration which is quite outstanding. As to maximum speed, this is about 80 m.p.h., and the adjustable steering wheel requires less than two turns from lock to lock. All these characteristics, and many lesser but, nevertheless, satisfactory ones, were observed and appreciated during the opening 63 miles.

Driving in Shifts

After dinner, it was decided to drive through the night in three-hour shifts. The North Road seemed comparatively deserted and a surprising sight was a tandem tricycle going great guns down Alconbury Hill. A little farther on, two solemn-looking members of the R.A.F. on bicycles pointed dramatically at the front end of the Midget as we swept past them. Thinking that the number plate must have fallen off or the fog lamp had come on, we carried out an investigation, but found nothing wrong and were left to suppose that this might be a new form of practical joke which certainly, in our case, achieved the desired results. While we were thus engaged a limousine Rolls-Royce, chauffeur driven and full of opulent-looking passengers, shot by at what we estimated must be a cool 80. Very little sound other than the swish of the tyres on the wet road and the positive blast of air denoted its passage.

It was getting dark as we reached Doncaster and nothing worth recording occurred until we reached Carlisle at 2.30 a.m. By this time the speedometer recorded 306 miles, and we had taken the trouble to ascertain that there was an all-night filling station available at this point. The M.G.'s petrol tank holds 13½ gallons, and this is a feature which should be praised from the house tops and copied by as many manufacturers of sporting cars as possible. It means that night journeys of great duration can be made with an easy mind. There may be several enterprising garages who remain open all night between London and Inverness, but we saw no sign of anything of the sort once it

NOTEWORTHY

THE **BORDER OIL CO., LTD.** LOWTHER STREET CARLISLE	Petrol and oil all night. Pumps under cover, and courtesy, even at 2.30 a.m.
THE GEORGE HOTEL BUCKDEN HUNTS	Well-cooked food, nice old furniture. Plenty of flowers and magazines. Covered parking space.
BALLACHULISH HOTEL ARGYLL	Highland hospitality at its best. Lovely position and most comfortable.
NETHY BRIDGE HOTEL INVERNESS-SHIRE	Large; very well run and designed for people who are prepared to pay a little extra in return for the best of everything.
FORTINGALL HOTEL GLENLYON PERTHSHIRE	They take their food seriously and the cooking is superb. No smoking in the dining room, and other good ideas.

grew dark, and the problem of refuelling on this and similar journeys will become a very great one when the petrol rationing is removed.

A breakfast halt was made at dawn near Abington. The Scotch Express toiling up the main-line gradient near there was the only other sign of life, and we certainly would not have exchanged the rows of first-class sleepers for our all-night journey in the M.G. Our intention being to proceed up the west side of Scotland, we then made our way towards Erskine Ferry. This method of crossing the Clyde is a popular one, but at 6 a.m. we were the first and only passengers of the day. The speedometer at this point read 415 miles and the weather began to deteriorate even more. By the time we reached the pass of Glencoe great gusts were

After an all-night run in appalling weather the Ballachulish Hotel provided an excellent breakfast in the true Scottish manner. The ferry across the loch is quite close to the hotel and operates fairly continuously.

Last Week-end in June—Contd.

screaming down at gale force and the rain hose-piped over the car, while all forms of life and livestock seemed to have gone for shelter rather than face such abnormal summer conditions It was, therefore, doubly pleasant to pull up at the hotel at Ballachulish. Unexpected and slightly bedraggled, we inquired about breakfast and were not only told with great politeness that a hot meal was ready there and then, but, best of all, we were advised that bathrooms and boiling water were at our disposal if we felt the need of them. This, in fact, was one of the things we had gone north to find out, quite apart from the technical requirements of the M.G. Co. There is a lot of criticism of British hotels, some of it, unfortunately, fully justified. On this journey, however, no meals were booked in advance, and at every stop, save one, the service and food provided were quite astonishingly good. The one exception was an English hotel which we called upon when returning south and which, from a lofty pre-war standard, has degenerated into a noisy and even noisome pot house not suitable for the traveller, nor desirable to a nation bent on developing its tourist industry.

Time to Spare

Once across the Ballachulish ferry (7s. for the M.G. and two passengers) we digressed considerably, calling on hospitable friends for lunch and hill hunting extensively in the Caledonian canal area during the afternoon. However, we arrived at Inverness in time for tea, and struck south again to the large and comfortable Nethy Bridge Hotel for the night. The speedometer had recorded, by this time, 615 miles. Half an hour was spent compiling an answer to a technical questionnaire prepared before leaving London. Sufficient to say that at this point it was not possible to add any oil at all to the engine nor, despite some appalling roads in the afternoon and a consistent main-road cruising speed in the neighbourhood of 65 m.p.h., were there any discernible rattles, squeaks or mechanical maladies. Occasional bursts of sunshine during the afternoon gave promise of a fine evening and at 7 o'clock the sun was beating down on the hotel verandah so that people sat about in the Continental manner and drank their pre-dinner cocktails out of doors. Unfortunately, the following morning produced more driving rain, but it was still possible to enjoy leaving the main road at Dalnacardoch and striking out over the mountain tracks to see what the war had done to once familiar moors and villages.

We had very little right to expect lunch from the hotel at Fortingall because we arrived late and without any warning, but they, nevertheless, took us in and fed us in a manner which did the greatest credit to the proprietor, who is a member of the Wine and Food Society. From this remote and attractive spot we pressed south after lunch, driving now really hard and using the gearbox to get the utmost out of the car. At Carlisle the friendly petrol attendant inquired about our adventures since he had seen us the previous day. The M.G. tank was filled up to the brim and the miles flew by as we pressed through bad weather towards the south. A stop for a brief meal and then on again, with the average rising all the time due to the absence of

Looking across Loch Ness from Glendoe Hill. The road on this side of the Loch is picturesque but in rather a poor state. The main road on the opposite side leading to Inverness is, however, in very good shape and allows excellent averages to be achieved.

lorry traffic on a Saturday night and generally deserted roads. Near London a building blazing at the side of the road caused police to divert all traffic round a loop-way, but in other respects the journey was uneventful and soon after midnight, the speedometer recording 1,161 miles, we drove into London well satisfied with the 1946 M.G.

Looking back, one was impressed by the fact that it was possible to enjoy so greatly a journey which might sound strenuous and could have been uncomfortable in such bad weather. From the touring point of view, the general excellence of the main roads, despite the war, and the splendid hospitality of the Scotch hotels are things to record with great pleasure. From the driver's point of view the road-holding, steering, braking and vivid performance of this latest addition to the long line of M.G.s means that in every way the car is right for fast touring. There was a time when a man who needed high performance and fresh air at the same time was catered for by a fairly wide choice of vehicles. To-day their number is narrowing, and such practical fitments as the external radiator filler, knock-off hub caps, revolution counter, folding screen, and really adequate petrol tank are almost things of the past. How thankful I am that a great tradition in small high-performance cars remains in such thoroughly sound hands as those of the M.G. Car Co.

The TC-type M.G. Midget.

As You Like It

"THE BLOWER," back from five years of Army vehicles, finds the 1946 M.G. Midget offers motoring—

MOTORING (or, more often, being motored) in Army vehicles is all very well—but it doesn't go far enough: or, sometimes, it goes too far. Either way, there is a longing for the sort of motors one used to drive and, if one's war happens to have been spent in the Far East, for the sort of roads one used to travel in days of peace that could rightly be called Piping (which adjective will not be found on the same page of the dictionary as Austerity).

Thousands of us must have felt this way, but not so many can have been as lucky as I was—lucky enough to be able to take up real motoring again just a little in advance of the point where it left off. There was rather the feeling of "This is where we came in" as I motored myself away in one of the latest TC-type M.G. Midgets—but with this difference; the setting seemed to have changed to gorgeous Technicolor.

There is just that difference between the post-war Abingdon product and the pre-war TA-type which took me to so many sporting events of various kinds up and down country. Extra sparkle was added to the TA in the introduction, just before Hitler put the lights out in Europe, of the TB model; and the TC has carried the process a stage further still.

Like the two models before it, the TC retains a dual personality. I struck the second side of its character first. Having been nurtured in the big wooffly Staff car tradition for five years, I had more or less forgotten what real sports car motoring can be. I slipped the Midget into gear, oozed off the mark, failed to remember that there was such a thing as a rev. counter which would be happily dancing round the top end

of its scale, changed up at a leisurely American speed and rapidly found myself in top still oozing about the place.

IN BRIEF

Engine.—4-cyl., o.h.v., push rod, 66.5 mm. by 90 mm., 1,250 c.c., 54.4 b.h.p. at 5,200 r.p.m. Tax £13 15s. Pump-assisted, thermostatically controlled cooling. Two S.U. semi-downdraught carburetters. S.U. petrol pump.

Transmission.—Borg and Beck single-plate dry clutch. Hardy Spicer open propeller shaft to spiral-bevel rear axle. Gears, 5.125, 6.92, 10.0 (synchromesh) and 17.32 to 1.

General.—Semi-elliptic springs all round. 13½-gallon rear tank. Weight (dry), 15.5 cwt.; length, 11 ft. 7½ ins.; width, 4 ft. 8 ins.; wheelbase, 7 ft. 10 ins.; track, 3 ft. 9 ins. Tyres. 4.50 by 19. Electrical equipment, Lucas-c.v.c. 12-volt.

Price.—£375 plus £104 18s. 4d. purchase tax (total, £479 18s. 4d.).

The odd thing was that I continued to motor in that sort of way for quite a while, using the top gear performance to no mean extent and forgetting all the tunes that can be played on the gear lever of such cars as the Midget and what happens in the playing.

Quite what reminded me that this wasn't really the best way to get the best out of something made in Abingdon, I forget; and it is quite immaterial. All that matters is that it suddenly burst on me that the stubby little gear lever was just asking to be used in the old way. So the shortest way from A to B was forsaken, and off we went down the first little lane that we came to and started playing tunes on the gear lever. That, perhaps, is not the happiest expression, because even a ham-fisted driver can't make much noise with the lever itself now that second gear has come into line with third and top with a synchromesh

change. The gears themselves are quiet, anyway. This useful addition in the TC has speeded up the transition from bottom to second to a straight-through movement of the lever performed without any qualms regarding crown wheels and pinions and the effect thereon.

The result of this improvement on the getaway of a car which turns the scale at only 15½ cwt. (dry weight) and possesses an engine pushing out 54.4 b.h.p. at 5,200 r.p.m. is obvious.

Taken up to its peak speeds in all the indirect gears, the Midget offers 24 m.p.h., 42 m.p.h. and 61 m.p.h. in first, second and third respectively and will slightly exceed these speeds if there is any real point in doing so. In top gear, the maximum appears to be so close to 80 m.p.h. as to make no matter and although (with no Brooklands or Donington available) I had no chance to make stop-watch tests, both experience on the road and slide rule excursions into theory, lead to ready acceptance of the M.G. experimental department's stop-watch figure of 78.26 m.p.h. as a mean maximum for a model tested with screen erected.

Maximum speed, however, is really of far less importance than top-end urge, and in this respect the new TC is far and away superior to the earlier T types. A desire for more when cruising at "60" or "65" (quite happy speeds in this car) is fulfilled, satisfyingly and without hesitation, by a little extra weight on the right foot. As a passenger so aptly remarked, "The accelerator works all the way."

These performance characteristics of the new TC are accompanied by all the things that should go with them in a car of this type. The new Luvax-Girling hydraulic dampers, in conjunction with the modified spring layout (employing rubber-bushed shackles in

place of the old sliding trunnions) improve both comfort and stability, the Lockheed brakes do all that is asked of them in the unobtrusive way that Lockheeds do, the Bishop cam steering is high enough geared to give lightning correction of skids (if they occur, which is not often), but is light enough for anyone to cope with without complaint and the column is adjustable for both rake and length—the latter by the popular Bluemel system.

Couple with all the above a body possessed of the usual attractive M.G. appearance, a fold-flat screen, a facia panel dominated by a pair of large dials for the speedometer and rev. counter (with the remaining dials and switches neatly panelized between), adjustable seats giving the ideal position for alertness and comfort and add a hood that isn't visible when it isn't wanted and offers, in conjunction with the improved rigid side screens, rather more than saloon snugness when it is—couple all these things together and you have some idea of the excellence of the TC body . . but with one most important fact omitted. The TC body is 4 ins. wider than its forerunners.

I started off by saying I was lucky by getting back to civilian motoring in such a car. On second thoughts, I'm not so sure. M.G.s wanted the car back, you see.

Engine room of the TC-type M.G. Midget, viewed from the near side. Twin S.U. carburetters can be seen at the back, oil cleaner low down to the right of the dynamo.

GARDNER ATTEMPT

Lieut.-Col. A. T. G. Gardner with his streamlined M.G. 750 c.c. record-breaker which, with 1,100 c.c. engine exceeded 200 m.p.h. before the war.

THE TRIALS . . .

From "Grande Vitesse," Ostend, Monday, 8 a.m.

BY the time you read this I hope that Lt.-Col. Gardner and his 750 c.c. six-cylinder M.G. will have captured the Class H records for the flying mile and kilometre, held since before the war by R. Kohlrausch in the four-cylinder Magic Midget.

Gardner's first attempt was timed for dawn yesterday, but when the party paraded it was raining dismally, and the weather continued bad throughout Sunday. As I write this the rain has ceased, and the high wind which has been blowing ever since I left England shows signs of dropping. With any luck, it should be possible for the M.G. to make a couple of trial runs later in the day and go out for the 150 m.p.h. speeds which Col Gardner expects to attain.

The M.G. is exactly the same as when Gardner took it to Italy in August to try for these records on the Brescia-Bergamo autostrada, except that a separate oil pump now looks after the lubrication of the Shorrocks blower and that the latter device has now been thoroughly tested—which was not so before. Even so, during the Italian attempt the car came within a fraction of a second of the record speed on one run, showing that 150 m.p.h. is well within the capabilities of the car.

In point of fact, Gardner hopes to reach his required speed at 7,600 r.p.m., whereas the engine will, of course run up to fantastic speeds around 8,000-9,000 r.p.m., or even more on the present gear ratio of 4.3 to 1.

The course lies some 15 miles out of Ostend in the general direction of Bruges and Ghent. The twin-track motor road, on which the measured sections were already marked out yesterday, runs east to west for some 25 miles, starting from a field near the remote village of Jabeke and running in almost a ruler-straight line to the even more remote hamlet of Aeltre, ending just short of the houses, again in a field. The road is reached off narrow side-lanes and is not officially opened to traffic, although a few cars and sundry farm carts use it from time to time.

This 25-mile strip of concrete, almost dead flat and with two or three very gentle curves, was constructed just before the war as the Belgian section of the projected London-Ostend-Istanbul trans-European highway. It is almost an exact replica of a German autobahn, with twin tracks separated by a grass strip on which grow bushes and saplings; fly-over crossings combine with the most modern bridgework and cloverleaf approaches.

Roughly in the middle of the 25 miles there is a 9-kilometre section (approximately 5½ miles) of dead straight, in the middle of which kilometre, mile and 5-kilometre sections have been marked off where the road is level. Gardner intends to use an over-all run of 15 kiloms. in which to get up maximum speed on the run-in and to lose velocity at the far end.

A photograph of the road appears on page 269.

AND THE CAR

THE remarkable speeds it is hoped to reach in Belgium to increase the Class H record by a large margin will be achieved on a car of somewhat mixed ancestry. The whole of the chassis and running gear was built up through the winter of 1933-4 for use with a standard K-type Magnette engine which, with Power Plus blower, developed 118 b.h.p.

The possibility of achieving 140 m.p.h. made a headline subject, and to assist in obtaining this speed the transmission was set at an angle with an offset crownwheel and pinion. This permitted a low seating position and reduced frontal area. Then, known as the Magic Magnette and fitted with a rather bulbous body with chocolate and yellow stripes, the car broke its class F record with a speed of 128.7 m.p.h., driver George Eyston, in October, 1934

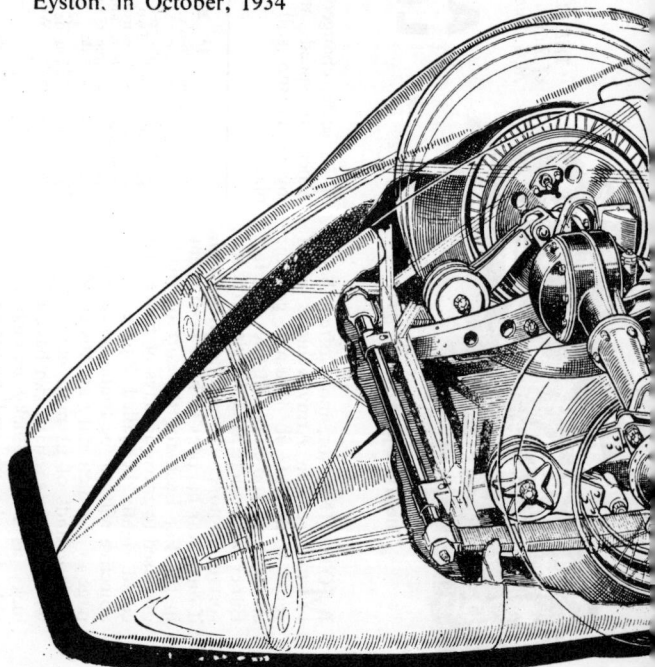

NEW SMALL CAR RECORDS

Details of the 750 c.c. Six-cylinder M.G. With Which He Hopes for 150 m.p.h. and More on a Section of Autoroute near Ostende

The engine of the 750 c.c. record breaker is also derived from the K type, although only a very few parts are now in accordance with the original design.

When Gardner began his record-breaking career it was with the narrow-body ex-Horton car which had been specially fitted with a Barronia cylinder head, Zoller type 5 compressor and various modifications to manifolding and valve gear by Robin Jackson. This engine gave approximately 170 b.h.p., and with it Gardner achieved 148.8 m.p.h. to gain the 1,100 c.c. record for a flying kilometre at Frankfurt in October, 1937. It was then suggested to him that he would be well advised to have a really low-drag body completely enclosing the wheels, but there were certain disadvantages in marrying such a shape to the standard chassis. To meet this problem the Jackson modified engine was removed and placed in the ex-Eyston chassis, the present body then being designed by Reid Railton. An additional 28 h.p. was secured by using a large Centric supercharger giving higher boost pressure, and in this form the car achieved 186.6 m.p.h. at Frankfurt in November, 1938, followed by 203.5 m.p.h. at Dessau in May, 1939, in both cases with 1,100 c.c., over the flying kilometre.

These figures were described by the chief of the racing department of Auto Union as the finest technical achievement in motor record breaking history, and there being no evident challenger to them, a decision was made to convert the engine to attack Class H 750 c.c. records.

In so doing the cylinder head and induction system arrangements have remained unchanged, but in place of the Centric supercharger of 4.68 litres per revolution

swept volume running at 0.478 engine speed, a new design, the Clyde, is employed with swept volume of 4 litres per revolution, running at 0.392 engine speed. This leaves the net supercharge, allowing for volumetric losses, unchanged at 28 lb per sq. in boost. The cylinder block, crankshaft, connecting-rods and four-ring Aerolite pistons are all specially designed for this effort. A variety of axle ratios is available, and with 41.1 fitted 150 m.p.h. equals 6,800 r.p.m. At this speed the total drag is equivalent to 80 b.h.p., and only 204 b m.e.p. is required, compared with the 340 b.m.e.p. realized at the same engine revolutions on the previous, larger, engine.

The dimensions are now 53 by 56 mm., compared to 57 by 71 mm. heretofore, the respective capacities being, therefore, 741 c.c. and 1,086 c.c. and the piston areas 20.3 sq. ins. and 23 7 sq. ins. In consequence of the drastic reduction in stroke a piston speed of 3,260 ft. per minute, which was the maximum achieved at the peak engine speed of 7,000 r.p.m. on the larger engine, is now the equivalent of 8,900 r.p.m. The engine in its latest guise has exceeded this speed with the power curve still rising. This marked increase of r.p m. has made it necessary to increase the valve spring pressures from 93 lb. to 129 lb. with the inlet valve fully open, and 90.5 lb. to 157 lb. with the exhaust valve fully open. Another minor change has been the substitution of Lodge sparking plugs type RL51 in place of the Bosch " 500 " plugs previously employed. Despite the reduction in nominal swept volume the two S.U. carburetters of $1\frac{7}{8}$ in. diameter have been replaced by instruments of the same make but $2\frac{3}{16}$ ins. diameter.

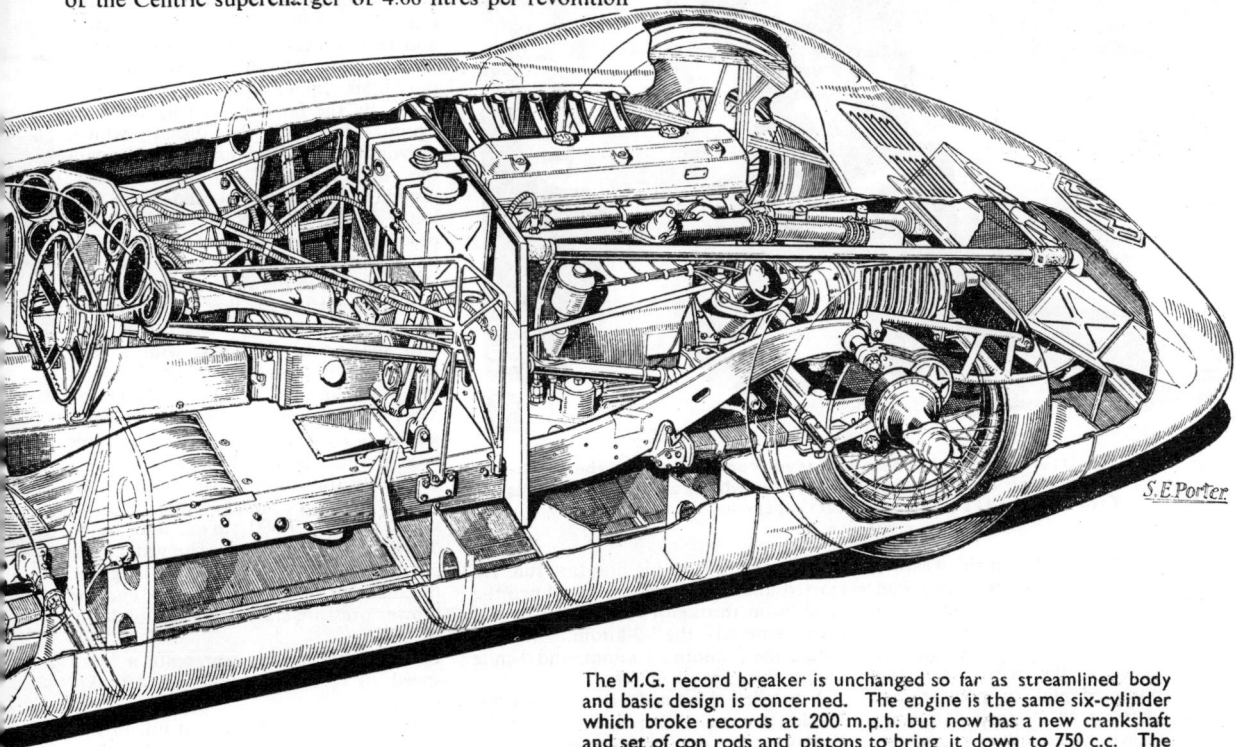

S. E. Porter

The M.G. record breaker is unchanged so far as streamlined body and basic design is concerned. The engine is the same six-cylinder which broke records at 200 m.p.h. but now has a new crankshaft and set of con rods and pistons to bring it down to 750 c.c. The supercharger is now a Shorrocks design, the Clyde, blowing at 28 lb. per sq. in. The engine develops 170 b.h.p. at 8,900 r.p.m.

GARDNER REACHES 164.72 m.p.h

ONCE again it has been left to the personal initiative of a private individual, Lieut.-Col. A. T. Goldie Gardner, to demonstrate to the world that British automobile engineering excels in the sphere of record breaking. Not only does Britain claim the fastest car on earth, but now she claims the fastest small car in the world—the six-cylinder 750 c.c. supercharged M.G., which set up new figures for the flying mile and kilometre at a shade under 160 m.p.h. last Wednesday, October 30, in the International Class H, subject to the usual official confirmation of the F.I.A.

This marvellously streamlined little 18-cwt. car now has the unique distinction of holding the highest speeds in three International categories—Class F (1,500 c.c.) and Class G (1,100 c.c.) at over 200 m.p.h., and now in Class H (750 c.c.) at over 159 m.p.h.—nearly 20 m.p.h. faster than the previous records.

Using a new supercharger designed by Mr. Chris Shorrocks, Gardner took the records with plenty of power in hand. His highest speed was about 172 m.p.h. as he flashed out of the measured mile on his fastest timed run of 164 m.p.h. average. This performance on the as yet unused motor road near Ostend, is a magnificent achievement which cannot fail to resound throughout the world, and Gardner, his team of mechanics and all those who contributed to this astonishing result deserve the congratulations of the entire British motor industry.

OSTEND,
Thursday, October 31.

YESTERDAY, the weather at last relented and a fine day awaited the record party when it arrived soon after dawn on the Jabbeke-Aeltre motorway. By 10 a.m. the early morning mist had risen and the road was dry enough for the first warming-up run. During the fag-end of the day

GETTING READY.—The M.G. has been unloaded from the lorry at the starting point. "Dunlop Mac" makes a detail check of the wheels and tyres while Les. Kesterton of S.U's (in cap and coat) supervises the filling up.

before, when it looked as if the weather would hold long enough, Gardner had done a warming-up run. He then changed plugs and set off in a fine rain to try a fast run up the course, but the mixture was on the weak side, the engine spluttered and to prevent any possible damage he stopped half way. The rain then came down in earnest, and the party called it a day. They also called it a certain sort of day.

Yesterday morning, however, was perfect. The carburation had been checked and slightly richened, and everything was ready when the sun came up. There was a slight wind which blew three-quarters ahead on the outward and three-quarters behind on the inward run. About 10.15 the hard plugs were in, the 45 lb. tyre pressures were checked and, without any fuss,

Gardner set off on his first fast run. He started about 15 kiloms. away down the road, went through two fast curves, and then entered the 9-kilom. straight, where the 5-kilom., 1-kilom. and 1-mile sectors were marked off.

I watched the run from a bridge just inside the measured mile. Away in the distance, on the horizon, where the ruler-straight road met the sky, a dot twinkled in the low rays of the sun. A

shrill whine sang down wind. The dot grew and grew until one could see the shape of the car, the green paint glistening and twinkling as it shot towards us, the wind slightly behind and to one side. It whined onwards rock steady, plumb in the middle of the road, just like a line drawn with pencil down the edge of a ruler. With a rush and a high-pitched shriek like an aeroplane, it shot the bridge and by the time I had whirled

round it was already a quarter of a mile away, dwindling into the distance until it disappeared on the other horizon and the whining note died and vanished. It was truly a magnificent, an impressive, and strangely exciting spectacle.

172 on the Clock

At the top end Gardner had the plugs checked again and one was changed, for once or twice the engine had stammered, and he had been forced to lift his foot. He had gone into the measured mile at only 6,600 r.p.m. and went up to 7,600 r.p.m. by the far end. Then, suddenly, again we heard the distant whine, growing louder and louder, and again the dot appeared in the distance, growing larger every second down that long, dead-straight road. This time I descended to ground level and crouched on the centre strip between the two carriageways in the shelter of the bridge. One moment the car was a green blur a mile away, the next it seemed to congeal into shape and shot past with a roar under the bridge which sounded like an explosion and he was gone, dwindling away once more into the far distance, going . . . going . . . gone . . . and this time, despite the half-head wind, he was obviously travelling much, much faster.

On this run he streaked into the measured mile at 7,600 r.p.m., and sailed down it at 8,200 r.p.m. The water temperature went up only to 70 degrees, the oil went up to just over 60, and the blower pressure stayed this side of 30 lb. of boost.

The 8,200 r.p.m. represents a road speed of 172 m.p.h. on the 4.3 axle ratio.

Thus, without any fuss at all, in two runs, the fastest 750 c.c. car in the world netted three more Class H records for Britain, and the attempt was over.

ITH 741 c.c.

Harry Hurkeyns, the Dutch M.G. driver, then attempted standing-start mile and kilometre records with his 12-year-old Marshall-blown K3 M.G. Magnette, but failed by about 3 secs., when his duralumin exhaust manifold and outlet pipes blew into pieces.

When I left Ostend, Enever and Jackson, Gardner's mechanics, were preparing the engine to attack 500 c.c. records on Saturday or Sunday, November 2 and 3, by means of paralysing two of the six cylinders.

THE SPEEDS
International Class H (750 c.c.)
(Subject to Official Confirmation)

FLYING KILOMETRE

1st run	..	14.54 secs.	. 247.592 k.p.h.	.. 153.846	m.p.h.
2nd run	..	13.58 secs.	. 265.095 k.p.h.	. 164.722	m.p.h.
Mean speed	..	14.06 secs.	.. 256.045 k.p.h.	.. **159.098**	**m.p.h.**

FLYING MILE

1st run	..	23.23 secs.	. 249.298 k.p.h.	.. 154.907	m.p.h.
2nd run	..	22.01 secs.	.. 263.226 k.p.h.	. 163.562	m.p.h
Mean speed	..	22.62 secs	. 256.128 k.p.h.	. **159.151**	**m.p.h.**

FIVE KILOMETRES

1st run	..	1 min. 17.04 secs.	. 233.644 k.p.h.	.. 145.179	m.p.h.
2nd run	..	1 min. 11.63 secs.	. 251.291 k.p.h.	. 156.144	m.p.h.
Mean speed	..	1 min. 14.33 secs.	. 242.147 k.p.h.	. **150.467**	**m.p.h.**

Flying Kilometre, Mile and Five Kilometre Class "H" Records all increased to over 150 m.p.h. in successful record bid on Belgian Autoroute near Ostend

THE RECORD RUN.—This fine Lane impression shows the M.G. entering the measured mile at 7,600 r.p.m. on its fastest run, when a speed of 164.722 m.p.h. was attained, the car, rock-steady, with one wheel outside the centre black strip. Inset is Lieut.-Col. A. T. G. Gardner, first post-war record man.

The Autocar ROAD IMPRESSIONS of 1946 Cars

TYPE TC M.G. MIDGET TWO-SEATER

TO anyone who has closely followed the development of the M.G. Midget from the earliest model away back in 1930 and who, furthermore, has been almost completely cut off from sports cars during the war, it is a most interesting and refreshing experience to renew acquaintance with the latest Type TC Midget. One of the first impressions received in driving it is that, as viewed today, the famous Midget has grown in stature; for one thing, today there are fewer sports cars available than in the past, secondly, the current model, with an engine of 1,250 c.c., is a far more substantial car than the earlier versions, which started with an engine of 847 c.c. and ranged in the course of time through intermediate capacities up to the present figure.

The great point about this car is, naturally, that it is intended for the section of motorists known as enthusiasts, but the appeal is far from being limited to those of ages in the twenties. The sports car characteristics which it embodies are appreciated by anyone who revels in driving for its own sake and who values accuracy of control and road stability developed to a fine art, allied with a performance that permits the car to be driven about as fast as anyone could wish under average conditions on main road or by-road, not to mention rough stuff of trials character.

It is a feature automatically derived from the whole design and layout of a car such as this that a driver strange to the M.G. feels at home in it at once and confident within a very few miles in driving it fast. Much is contained within the bare citation of that fact. It means that in the first place the driving position is right; that driving vision, with full view of both wings, is as it should be; that the controls are correctly placed; that the car can be cornered without conscious steering effort; that its suspension holds it firmly to the road, and that the brakes behave as they should.

As is well known, the type of motorist whose natural choice of car is the family saloon would be critical of some of the features that belong to such a car. In other words, with the Midget there is noise from the engine, though not such as to be tiresome or wearying, and the suspension that gives fine roadholding is hard, especially at the lower speeds. To omit reference to this aspect would be not to give a complete picture of the car, but they are not disadvantages to deter the enthusiast.

Under present conditions it has not been possible to record the range of performance data, but interesting as such details can prove for comparison there are other broader factors of a car's performance which can weigh even more in passing judgment. Especially to the experienced motorist it will mean much to state that during a test totalling approximately 470 miles the Midget covered at night 90 miles of main road, measured by the speedometer, in two minutes less than two hours, non-stop. It is interesting that in the first hour 45 miles were covered. The route in question is a good one for safe, fast driving, and the weather was dry and clear. At other times during the test it proved possible to achieve that yardstick of soberly verified average-speed performance, namely a solid 40 miles in the hour, without the driver putting forward exceptional effort or taking more out of the engine than seemed reasonable. A highest speedometer reading of 80 was seen on one occasion, 75 once or twice and 70 several times. Approximately, 1,000 r.p.m. on top gear equals 15 m.p.h. As an extreme reading the speedometer needle can be pushed round close to the 60 mark on third and towards 40 on second gear, and 50 and 30 can be recorded as comfortable readings on those gears.

The engine, even when pressed by the driver, does not become rough or harsh, but has about it a feeling of mechanical hardness, not easily defined, which somehow is satisfying in a sports engine. It is not meant to trickle along on top gear, though it is sufficiently flexible to below 20 m.p.h. on top. There was no more than a trace of pinking even on the present petrol.

The remote control gear change, with its short, firm lever, is extremely satisfactory. Second, third and top have good synchromesh, but in the main the double-declutching process is naturally employed for fast downward changes with a car such as this. The changes go through beautifully smoothly, and quickly enough on

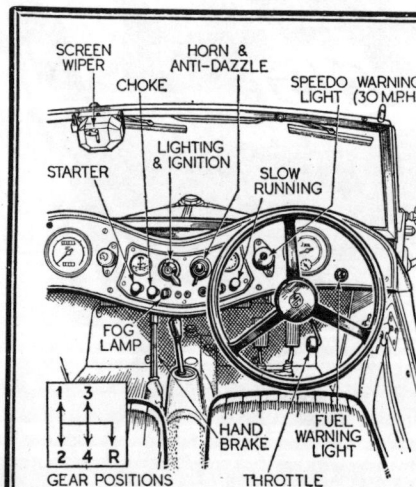

DATA FOR THE DRIVER

TYPE TC M.G. MIDGET

PRICE, with open two-seater body, £412 10s plus purchase tax, £115 6s 8d, total £527 16s 8d.

RATING: 10.97 h.p., four cylinders, o.h.v., 66.5 x 90 mm., 1,250 c.c. Tax, £13 15s (cars first registered after January 1, 1947, tax £13).

WEIGHT, without passengers: 15 cwt 3 qr. **LB. PER C.C.**: 1.41.

OVERALL GEAR RATIOS: 5.125, 6.93, 10.00 and 17.32 to 1.

TYRE SIZE: 4.50 x 19in on knock-off wide-base wire wheels.

LIGHTING SET: 12-volt. Automatic voltage control.

TANK CAPACITY: 13½ gallons; approximate fuel consumption, 29-35 m.p.g.

TURNING CIRCLE: 37ft. (L. and R.).

MINIMUM GROUND CLEARANCE: 6in.

MAIN DIMENSIONS: Wheelbase, 7ft 10in. Track (front and rear), 3ft 9in. Overall length, 11ft 7½in; width, 4ft 8in; height, 4ft 5in.

Diagram labels: SCREEN WIPER, HORN & ANTI-DAZZLE, CHOKE, SPEEDO WARNING LIGHT (30 M.P.H.), LIGHTING & IGNITION, STARTER, SLOW RUNNING, FOG LAMP, GEAR POSITIONS (1 3 / 2 4 R), HAND BRAKE, FUEL WARNING LIGHT, THROTTLE

Described in "The Autocar" of October 12, 1945.

the upward movements for good acceleration results. Smooth starting from rest is obtained without special care in engaging the clutch.

For these days the steering is quite high geared—about 1¾ turns of the wheel give the full lock-to-lock movement. At no time is the steering conspicuously heavy, and it is admirable, in conjunction with the suspension characteristics, in allowing the car to be placed just where the driver wants to aim it. Corners of fast or slow type can be taken in a smooth, clean sweep on a selected path for the wheels and, as compared with some more modern and more comfortable suspension systems, the driver knows where he is with the car. That is, if he overdoes things when cornering he encounters tyre scream, and at a further stage a tail slide to act as a warning, and can pull out of the difficulty he has created for himself. The Lockheed hydraulic brakes reduce speed just as one would wish, and do not by any fierceness of action suggest that they are doing as much work as in fact they are frequently achieving in rapidly slowing a car of this description.

FRONT TRACK 3' 9"   WHEELBASE 7' 10"   REAR TRACK 3' 9"

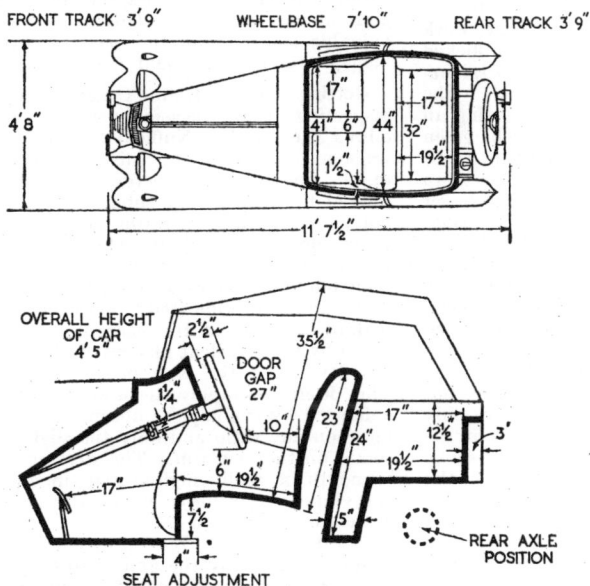

Measurements are taken with the driving seat at the central position of fore and aft adjustment. The diagrams are to scale.

Driving position is of the kind again appreciated by the keen driver, with a back rest which can be adjusted close to the vertical, and the spring-spoked wheel in a position where one has full power over it. Adjustments in connection with driving position are well arranged; both the column rake and the wheel itself can be adjusted, and the seat cushion is easily moved fore and aft. It is most satisfactory, too, to sample again the fly-off type of hand brake lever, held to its ratchet only when the knob is depressed, and released by a slight pull.

It takes a little while to become used to horn and antidazzle controls on the instrument board instead of at the centre of the wheel. The horn is sufficiently powerful. A definitely inconvenient feature is that there is nowhere for the driver comfortably to place the left foot for long periods except to rest it lightly on the clutch pedal. It would seem that, with advantage, the clutch and brake pedals could be moved a little to the right, thus also rendering heel-and-toe braking change-downs easier for those who sometimes use this method. The passenger has ample leg room. Another criticism is that apart from ordinary door

pockets and the main luggage space behind the front seats, where a useful quantity of baggage can be put under cover, there is no shelf or cubbyhole for the oddments that one always carries.

A good general view is given by the externally mounted mirror, but with hood and side screens raised the mirror view of vehicles close astern is limited. The hood is easily put up and down and the protection afforded, in conjunction with the easily erected side screens, is more than adequate.

Points about the engine compartment are the general neatness, and an accessible oil filler and dipstick; the battery and tools are in convenient lidded metal containers. The engine started readily from cold, taking a little while to gain working temperature without use of the mixture control.

C'est Formidable

Continued from page 43

rugged coastline we had expected in Cornwall and the little fishing villages of Brittany. The bleak highland charm, the kilts, the plus fours, and the life-saving tingle of a glass of dark brown sherry will not be soon forgotten.

Thawing out began as we drove south to the boat, stopping en route at Abingdon-on-Thames, where we watched the factory assembling its 53 MG-TD's a day. The service manager at the works was a friendly man full of information. He told us the cost of an exchange for a factory rebuilt engine would be $112, including installation. Sorely tempted by this generous price, we at length decided to keep our old engine, and worry about troubles when they came.

They haven't come yet. My MG, nicknamed *C'est Formidable* somewhere in France with a brass plate on the dash to prove it, is sitting proudly in the garage at home, ready to carry us another 11,000 miles free of trouble.

Perhaps motoring isn't the best way to tour Europe, but you'll have to prove it to me. The strain of planning railway tickets in advance and rushing to meet schedules would take all the fun out of traveling. After paying $300 to carry the car both ways across the ocean, our only travel expense was gas and oil, and we were able to spend the nights outside of the large cities at ridiculously low rates. To any person touring Europe by car I recommend the *Continental Handbook* published by the British Royal Automobile Club.

For five months we beat the stuffing out of our MG—overloaded her, failed to change the oil or grease her properly, drove up mountains and down cliffs in boiling or freezing weather, and she just bobbed up for more. I enjoy traveling in a car like that.   **THE END**

THE M.G. "T.C." MIDGET 2-SEATER

I T is good to know that the ever-popular M.G. Midget has survived the war and is in production again. It is the same trim, efficient 2-seater we knew pre-strife times, with 1,250-c.c., push-rod, o.h.v. 4-cylinder engine, 4-speed gearbox, and ½-elliptic suspension, now improved in a number of practical ways and endowed with greater elbow-room and even better weather protection.

We took one of these cars over from the M.G. works at Abingdon last November and subjected it to a strenuous 400-miles test ; the more we drove it, the more reluctant did we become to take it back to Mr. Cox, M.G.'s Publicity Manager. From the commencement a driver feels at home in this M.G. and, as he enthuses over the comfortable driving position and the layout of the controls, his passenger is invariably praising the comfortable seating and the very generous leg room. The bench-type front seat has been contrived so that not only is it possible to slide the two separate cushions forward, but also to adjust the rake of the squab, while the steering column is telescopic, so that the best seating position is quickly attainable. The seat strikes just the right balance between sponginess and hardness, and, while perfectly comfortable, the driver is prevented from rolling about when indulging in faster-than-normal cornering. For competition work raising the cushion some 3 in. would aid visibility, as the bonnet slopes upward to some extent, although both wings are visible in any case. M.G.s have always fitted the near-perfect handbrake and the centrally-disposed lever on the "T.C." is no exception. Of "fly-off" type, it really holds the car, releases instantaneously when pulled back, and locks effectively if the thumb-catch is pressed down. It is indeed an excellent brake.

The central, remote-gear-lever calls for equal praise. It could hardly be better placed, is short, absolutely rigid, with a pleasantly slim grip. This is well merited, for the synchro-mesh gearbox is one of the nicest we have operated. The synchro-mesh works well, but double declutching is equally effective, and upward and downward changes are really quick. The change back into top gear from 3rd is very pleasant, helped by the sensible positioning of the lever. It is also quite practical to "snatch" upward changes with the throttle foot held down. The only care necessary is not to pull the lever too far to the right when going from 2nd into 3rd, or it tends to catch on the gate ; the reverse position works easily against a spring. The clutch is light and works well, but it might be a trifle more progressive. The brake pedal needs fair pressure, but gives excellent results.

The wood facia has a centre panel carrying Lucas ammeter, ignition and lamp switch, horn button and dipper, Jaeger oil gauge reading to 160 lb./sq. in., starter pull, mixture pull, fog-lamp switch, battery charging socket, panel light switch, and screw-type, slow-running adjustment. It is flanked by a dash-

A very well-appointed car with a lively performance. Excellent roadholding and braking, and effective weather-protection.

★

lamp and a 30 m.p.h. warning lamp and, on the left, is the 100 m.p.h. Jaeger speedometer with mileometer and trip, and on the right the Jaeger rev.-counter reading to 6,500 r.p.m. and having an inset clock. Also, on the extreme right, is a window which flashes the word "Fuel" when the tank capacity is down to about two gallons ; unfortunately, it does this very vividly, right in the driver's eyes. Speedometer and rev.-counter are simply but effectively calibrated. The oil gauge is somewhat blanked by the steering wheel, but not seriously ; the lamp dipper is rather too close to the wheel and could with advantage work the other way, so that it could be flicked with the left forefinger, instead of having to be fumbled for with the thumb. The panel lighting is adequate for reading all the instruments, but too much green light leaks round them for the light to be left on while driving. The starter is up to its task and the mixture-enricher for the twin S.U. carburetters is spring-loaded to obviate driving off in "rich"—a good point. The ignition key in the "off" position does *not* render the electrics dead, yet has to be "on" to work the wipers. The dynamo gives a good, controlled charge and the dual screen wipers are efficient but very noisy. Normal oil pressure is 42 lb./sq. in. and does not vary with hard driving. There are only two hexagon-motifs now visible from the seat, one in the centre of the 3-spoke steering wheel and one on the back of the licence holder. The scuttle has the two familiar wind-deflecting "humps" and the screen folds flat, with wiper box before the passenger. Entry into the M.G. is no more difficult than in any other low-built car of this type, and is aided by the low running boards. Door handles

and bonnet fasteners work effectively, and each door has a pocket. There is no cubby hole.

Our initial impression of the "T.C." Midget was its trim, well-balanced appearance and the high quality of the finish. The car could certainly take its place unashamedly with the "limousines and landaulettes" outside the best hotels, while there was practically no suggestion of austerity. There are carpets on the floor, and a wind excluder round the handbrake, and the finish of the car—red in our case with upholstery to match —and the equipment included, leave nothing to be desired. A most useful item of the body layout is the generous luggage space behind the seat. This is covered by a "tonneau"-cover when the hood is down, or by the hood when this is erected, and will hold several suit-cases, still with room to spare for coats and similar etceteras. This is a very valuable feature, and the need to carry luggage externally should never arise.

Getting away from Oxford, after inspecting that pleasingly-stark 1923 o.h.v. M.G. in the Nuffield Showrooms, we soon found we were cruising at 55-60 m.p.h. on the speedometer. The M.G. proved to have a very subdued exhaust note, even when accelerating in the lower gears, and in towns it attracted only favourable comment and attention. We soon found ourselves keeping the engine speed above an indicated 2,500 r.p.m. by making full use of the gearbox, encouraged by the excellent placing of the rigid gear-lever and the ease of the change. This desire to employ the lower ratios is enhanced by a complete absence of gear noise on any ratio and only the slightest sound on the over-run. The steering we found to be really high-geared (in quite the vintage tradition) ; it actually needs 1⅝th turns, lock to lock. The lock is moderate. As the test progressed we confirmed not only the ability to control the car by wrist-movements alone, but that no return motion is felt through the wheel on any surface, and that there is

The "T.C." M.G. Midget ready to undertake the MOTOR SPORT *timed tests.*

admirable castor action. The wheel judders in the hands at times, but never to an abnormal degree, the scuttle, like the radiator, being commendably rigid. It is rather heavy steering through appreciable arcs, but very reasonable in normal motoring. The door cut-away tends slightly to impede the elbow when " dicing." This steering is in no way " spongy," and after 7,500 miles showed little lost motion.

The M.G. corners as well as its predecessors. If anything, it understeers, which is all to the good, especially as the quick castor action brings it out of corners very nicely. We could not make the car slide, even on wet roads, and it steers accurately both on the straight and when cornering. The steering remains good when reversing, which is useful in special tests, and the car certainly does not roll, even under " rally-test " driving. The tyres protest rather early, but not too loudly. The suspension is pleasingly hard in quite the " old-school " manner, yet the car is not uncomfortable and can be taken over gulleys and bad surfaces without any feeling of remorse. This firmness of the suspension undoubtedly endows the " T.C." M.G. with the good roadholding aforementioned and, if it occasions a few body rattles, we feel these are entirely forgivable in view of the pleasant handling qualities achieved.

The brakes are really good. They call for fairly firm pressure on the pedal, but have a secure, hard feel, are very powerful, and progressive braking is quite easy to accomplish. There is only occasional brake noise and in the wet, if the wheels are allowed to lock, the car remains controllable.

On the first day of the test a cold wind made us resort to the excellent weather protection which we were so glad of later on. Four rigid sidescreens are stowed in a felt-lined locker at the back of the luggage compartment. They fit, two on each side, by inserting metal tongues into slots at the back and metal sockets over studs at the front, where wing nuts secure them. We soon erected the front screens and found that they excluded side draughts. When we encountered driving rain and a gale-force wind we erected the disappearing hood and the rear side screens, and in a matter of miles gave full marks to the weather protection of the new M.G. Midget. Hardly any rain drove in and the interior of the car was literally warmer than that of a saloon —so much so that in less severe conditions we should have removed the rear sidescreens in order to ventilate the car. The weather really was quite abnormal and the M.G. came through with flying colours, the interior almost bone dry. No one need have any qualms about using this car in winter, and this protection is rendered practical by two windows in the back of the hood, permitting of easy reversing, and by signalling flaps at the base of each sidescreen, normally secured by a press-stud tab.

As we have observed, the driving position is generally comfortable, but, unfortunately, there is nowhere to stow one's clutch foot and the accelerator is rather difficult to hold fully depressed. The

fuel tank holds the useful quantity of 13½ gallons and possesses a very excellent quick-action filler cap.

Two other good points about this M.G. deserve special emphasis. One is the provision of centre-lock wire wheels, rare on modern cars. They carry 4.50 in. by 19 in. Dunlop synthetic tyres. The other is the excellent lighting. It is possible to drive at maximum speed after dark, thanks to the long-range beams from the Lucas headlamps, yet these " dim " effectively, while the small Lucas spotlamp is one of the finest we have driven behind. The sidelamps rather reflect in the headlamp plating and so can be checked as " on " from the seat; there is a good brake lamp, but no reversing light.

M.G. MIDGET " T.C." 2-SEATER

Engine : 4 cylinders, 66.5 by 90 mm. (1,250 c.c.), 11 R.A.C. h.p.

Gear Ratios : 1st, 17.32 to 1 ; 2nd, 10.0 to 1 ; 3rd, 6.92 to 1 ; top, 5.125 to 1.

Tyres : 4.50 in. by 19 in. Dunlop synthetic.

Weight (in road trim with approx. 3 gall. of fuel, but less occupants): 16½ cwt.

Steering Ratio : 1⅝th turns, lock to lock.

Fuel Capacity : 13½ gall. (2-3 in reserve).

PERFORMANCE DATA

Acceleration :
0-50 m.p.h., 16.25 sec.⎱ mean of
0-60 m.p.h., 27.25 sec.⎰ two-way
                          runs
s.s. ¼ mile : 23.45 sec. (mean), 22.0 sec. (best run).

Speed :
f.s. ¼ mile : 63.50 m.p.h. (mean), 65.75 m.p.h. (best run).
Maximum in indirect gears (corrected for speedo. error) :
1st : 22.6 m.p.h. at 5,500 r.p.m.
2nd : 39.2 m.p.h.   ,,
3rd : 56.6 m.p.h.   ,,

Fuel Consumption : Approx. 27 m.p.g.

We subjected the car to our usual timed tests, and here it was decidedly unlucky. Wind was gusting up to gale force across the course, which was sodden with rain—*very unfavourable conditions.* The figures we obtained are given in the accompanying table, but before you commit them to memory, some explanation is necessary. At Brooklands it was possible to work a car up to its true maximum speed. Under prevailing conditions we cannot do this, but our speed for the flying ¼-mile is a fair approximation of what can be expected under road conditions. As usual, we did several runs in both directions of the course with the screen down, and carried a lightweight passenger and only a few gallons of petrol. The main figures were : flying ¼-mile at 63.5 m.p.h., mean speed, best run at

65.75 m.p.h. Best standing ¼-mile in 22 sec., 0-50 m.p.h. in 16.25 sec., 0-60 m.p.h. in 27.25 sec. The wet road affected the braking, which from 30 m.p.h. to a standstill occupied 41 ft ; the car slid with locked wheels and did well in the circumstances. Incidentally, before doing these tests we check the speedometer to eliminate inaccuracies ; weight is ascertained on the same weighbridge in each case.

The engine of the " T.C." M.G. is smooth and free from flat-spots, so that the speed attained before changing into a higher gear really rests with the driver. An indicated speed of 3,000 r.p.m. is very pleasant, or 4,000 r.p.m. for brisker occasions. The engine sounded to have reached its safe limit at an indicated 5,500 r.p.m., and the corrected road speeds were then 22½ m.p.h. in 1st gear, just over 39 m.p.h. in 2nd gear, and 56½ m.p.h. in 3rd gear. In top gear we got an indicated speed of 4,400 r.p.m. entering the measured stretch and 4,600 r.p.m. leaving it, on a run timed at 62.1 m.p.h., which shows that, given a longer run, the car would probably have improved on its maximum speed. On the road, in fact, on one occasion the speed exceeded 70 m.p.h.

It was most interesting to find that the engine did not protest in the slightest degree to " Pool " petrol. It started at once from cold, but needed some encouragement from the enrichener before it would pull. It always cut clean on the switch, and in just over 400 miles no oil or water was needed. The fuel consumption, checking a tankful against the trip reading, came out at 27 m.p.g., much town work and all the timed tests included. In top gear the engine began to " take hold " above an indicated 3,000 r.p.m., and in the same ratio would run down to a few m.p.h. without transmission snatch. Beneath the bonnet the oil filler is readily accessible on the valve cover, and coil, electric fuel pump and junction box equally so on the bulkhead. The external mirror is well placed, but visibility suffers with the sidescreens up on a wet night. The starting handle is clamped to the front of the luggage shelf and the tools are in a locker beside the battery box on the bulkhead. The engine is finished in light grey paint and the rev.-counter is driven from the belt-driven dynamo. The wheelbase is 7 ft. 10 in., the track 3 ft. 9 in., and those with garaging problems may like to know that the overall dimensions are 11 ft. 6 in. by 4 ft. 8 in.

To sum up, the "T.C." M.G. Midget is a good-looking, attractive car. It corners very well indeed, and its excellent gear-change and good brakes, together with " vintage-like " roadholding and suspension, enable it to live up to its slogan of " Safety Fast." Its other characteristics, if less outstanding, are equally satisfactory, and its economical speed, willing acceleration and very practical and complete equipment, combine to render this car a thoroughly useable 2-seater. Full details from the M.G. Car Company, Ltd., Abingdon-on-Thames, Berkshire. The price, with purchase tax, is £527 16s. 8d.—W. B.

The Motor Road Test No. 3/47—

Make: M.G. **Type:** Midget, Series TC.

Makers: M.G. Car Co., Ltd., Abingdon-on-Thames, Berks.

Dimensions and Seating

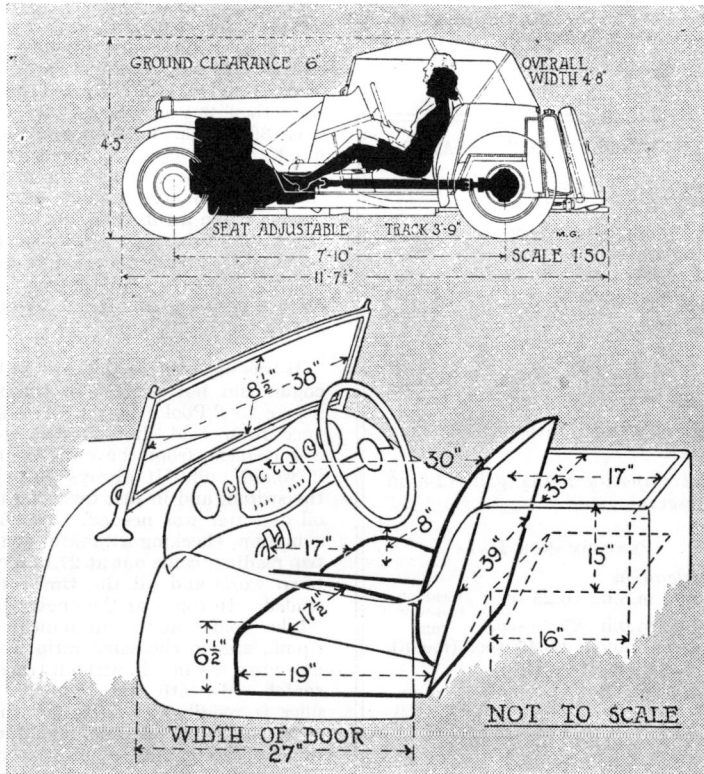

GROUND CLEARANCE 6" OVERALL WIDTH 4'8"

SEAT ADJUSTABLE TRACK 3'9"

7'10"

11'7¼" SCALE 1:50

8½"-38"

30"

35" 17"

17" 8"

39" 15"

17½" 16"

6½"

19"

WIDTH OF DOOR 27"

NOT TO SCALE

In Brief

Price £412 10s.
  plus Purchase Tax £115 6s. 8d. =
                          £527 16s. 8d.

Tax	£13.
Road weight unladen.. ..	16½ cwt.
Laden weight as tested ..	19½ cwt.
Fuel consumption	33 m.p.g.
Speed	72.9 m.p.h.

(mean both ways, screen up)

61 m.p.h. 3rd

42 m.p.h. 2nd

Acceleration .. 10-30 on top 11.9 secs.

0-50 through gears 13.9 secs.

Tapley pounds per ton and
  gradients : 195 lb.=1 in 11½ max. on top

285 lb.=1 in 8 max. on 3rd

400 lb.=1 in 56 max. on 2nd

Gearing .. 15.5 m.p.h. on top at 1,000
  r.p.m. 2,500 ft. per min.
  piston speed at 67 m.p.h.

Specification

Cubic capacity	.. ..	1,250 c.c.
Cylinders	.. ..	4
Valve position	.. ..	o.h.
Bore	.. ..	66.5 mm.
Stroke	.. ..	90 mm.
Compression ratio	.. ..	7.4
Max. power	.. ..	54.4 b.h.p.
at ..	.. ..	5,200 r.p.m.
H.P. per sq. in. piston area	..	2.5
B.h.p. per ton unladen		66
Piston area per ton unladen		26 sq. ins.
Litres per laden ton/mile	..	2,420
Ft. per min. piston speed at		
max. h.p.	.. ..	3,068
Carburetter	.. ..	Twin S.U.
Ignition	.. ..	Coil
Plugs	.. ..	Champion L.10S.
Fuel pump	.. ..	S.U.
Oil filter	.. ..	Full-flow
Clutch	.. ..	Borg and Beck single plate.
Top gear	.. ..	5.12
3rd gear	.. ..	6.93
2nd gear	.. ..	10
1st gear	.. ..	17.3
Reverse	.. ..	17.3
Propeller shaft	.. ..	Hardy-Spicer
Final drive	.. ..	Spiral bevel
Brakes	.. ..	Lockheed
Drums	.. ..	9 ins. dia.
Friction lining area	..	104 sq. ins.
Friction lining area per ton unladen		126 sq. ins.
Steering gear..	.. ..	Bishop cam
Tyre size	.. ..	4.50×19

Fully described in " The Motor," October 10, 1945.

Test Conditions

Dry concrete road, strong wind, Pool petrol.

Test Data

ACCELERATION TIMES on Two Upper Ratios

	Top	3rd
10—30 m.p.h.	11.9 secs.	7.9 secs.
20—40 m.p.h.	12.2 secs.	8.0 secs.
30—50 m.p.h.	13.0 secs.	9.3 secs.
40—60 m.p.h.	17.2 secs.	11.6 secs.

ACCELERATION TIMES Through Gears

0—30 m.p.h.	5.8 secs.
0—40 m.p.h.	9 secs.
0—50 m.p.h.	13.9 secs.
0—60 m.p.h.	21.1 secs.

MAXIMUM SPEED : Flying Quarter-mile

Mean of four opposite runs (hood and side screens erect)	72.9 m.p.h.
Fastest run	78.9 m.p.h.
Mean of four opposite runs (screen flat)	73.2 m.p.h.
Fastest run	77.6 m.p.h.

FUEL CONSUMPTION

Overall consumption for 372
  miles—33 m.p.g.
43.5 m.p.g. at constant 30 m.p.h.
39 m.p.g. at constant 40 m.p.h.
34 m.p.g. at constant 50 m.p.h.
28.5 m.p.g. at constant 60 m.p.h.

BRAKES at 30 m.p.h. (Tested on wet concrete road).

0.85 g.(=35.5 ft. stopping distance) with 90 lb. pedal pressure.
0.75 g.(=40 ft. stopping distance) with 50 lb. pedal pressure.
0.48 g.(=63 ft. stopping distance) with 25 lb. pedal pressure.

HILL CLIMBING

Maximum top-gear speed on 1 in 20	61 m.p.h.
Maximum top-gear speed on 1 in 15	53 m.p.h.

STEERING

1½ turns of steering wheel lock to lock'
Left- and right-hand lock—37 ft.

Maintenance

Fuel tank : 13½ gallons, including 2½ reserve. **Sump :** 9 pints. **Gearbox :** 1½ pints. **Rear axle :** 2 pints. **Radiator :** 14 pints. **Grease points :** 21 nipples, grease gun lubrication. **Spark timing:** T.D.C. Plug gap: 0.02 in. **Contact gap:** 0.01 in. **Tappets :** 0.019 in. (hot). **Front wheel toe-in :** 1½ ins. **Castor angle :** 8 degrees. **Damper fluid :** Luvax piston-type oil. **Tyre pressures :** 24 lb. **Oil filter element :** Change every 10,000 miles. **Battery :** 12-volt, 51 amp./hrs. **Head-lamp bulbs:** Lucas, type 52 and 54. **Side, rear and stop :** Type 207. **Trafficators :** Nil. **Map-reading light :** Lucas 207.

Ref. No. B. 13/47.

The 1947 M.G. TC-type Two-seater

A Sports Car in the Traditional Style

THE TC-type M.G. two-seater is essentially a sports car in the traditional style, and following a prolonged spell of sedate motoring in a family saloon of equivalent h.p., initial mileage on it necessitated a rapid readjustment of reactions and outlook.

This newest model in the T series introduced in 1946 bears a marked outward resemblance to its immediate predecessors the TA and TB models. The chassis is similar to the TA and the engine identical with the TB, i.e., it is a short-stroke, high r.p.m. unit; only when one examines the car more closely do the differences become apparent. In particular does this apply to the bodywork, which is now 4 ins. wider and thus provides greater elbow-room and freedom of movement.

In the course of a 600-mile test the M.G. was used in heavy traffic, on long main-road runs, and over country of the trials variety, and showed itself to be equally happy on every sort of going, with adequate ground clearance on deep ruts, and electrical and respiratory fittings located well up out of the wet.

Top-gear Flexibility

Initial pottering around and about London, once the general feel of the car had been acquired, showed docility and flexibility not usually associated with a car of this type. Although at its best at high r.p.m.—its peak is 5,200—the engine pulls along quite happily at 25 m.p.h. in top and accelerates steadily away with very little pinking, provided the accelerator is not depressed too violently; nor is there noticeable snatch in the transmission until the engine is throttled down almost to stalling point. On the other hand, the acceleration on the indirect speeds is such that most other cars can be left behind when the lights go green.

Intelligent use of the gearbox on the M.G. is particularly easy; the short gear lever is very conveniently placed, the movement is short and almost any sort of gear change from full use of the synchromesh, with which the three upper ratios are fitted, to the more brutal snatch through, can be made with complete ease and in absolute silence.

On the model tested there was a certain amount of whine on the indirect gears, particularly towards peak r.p.m., but this could not be considered obtrusive, and with the hood down was scarcely audible.

The clutch, too, is in keeping, being free from snatch on take-off, yet showing no indication of slip under the most violent treatment during acceleration tests through gears.

Suspension on the TC-type is completely orthodox with short semi-eliptics at front and rear controlled by Luvax piston-type shock absorbers. This combination, coupled with a rigid frame, provides a very solid feeling and results in good stability at speed. In fact, during the course of the test the extremely slight effect of a strong crosswind when the speedometer was registering well over 60 m.p.h. was noteworthy.

Short of fitting remote control shock absorbers, and one would scarcely expect to find these a standard fitment on what is, after all, the cheapest sports car on the market to-day, it is difficult with such orthodox springing layout to arrive at a suspension which gives truly satisfactory results throughout the whole range and over all types of going. The M.G. is patently designed to be driven faster than most and is sprung accordingly. Consequently, it must be admitted that at speeds below 40 m.p.h., except on roads with an impeccable surface, the ride becomes increasingly harsh as the speed diminishes, and this effect is enhanced by the slightness of the upholstery.

Adjustable to Suit

Generally speaking, the M.G. is a very controllable car. Lateral adjustment both for seat and steering column enables a sitting position to be achieved by people of varying heights which is just about right, and the controls, with the exception of the hand-brake lever, of the fly-off type, which, to our mind, is tucked too far away under the scuttle, are conveniently placed.

The steering is high geared—1½ turns of the wheel being necessary from lock to lock—and is, in consequence, not light. This fact is not particularly noticeable under normal conditions, but on fairly sharp turns and when manœuvring in confined spaces, more effort is necessary than one would normally expect in a car weighing only 16½ cwt. Furthermore, one feels that a smaller turning circle than 37 ft. would be an advantage, particularly for competition work.

On a vehicle of this type the ability to stop quickly is of particular importance, and in this

FAMILY LIKENESS.—The TC-type M.G. bears a strong resemblance to the long line of models which made their name in pre-war competition.

RESPIRATION. — The twin S.U. carburetters draw air through a large filter mounted, like all other engine auxiliaries, high up out of the way of road moisture.

regard the M.G. scores full marks. The brakes, without being abrupt, are really effective; in fact, there is one cyclist who, in the course of the test, chose a most inopportune moment to make a violent swerve towards the centre of the road, and undoubtedly owes his life to them. Application is light and at no time throughout the somewhat lengthy test, which involved, on occasion, some fairly drastic braking, was there any noticeable brake-fade.

Petrol consumption these days has a considerable bearing on one's motoring, and the 33 m.p.g. obtained with the TC is entirely praiseworthy, particularly as one normally credits a twin carburetter engine giving 54 b.h.p. from 1,250 c.c. with rather less than average economy. With this consumption and the tank capacity of 13 gallons the M.G. has the useful cruising range of 430 miles. The petrol recording device on the M.G. is unusual in that a green warning light is substituted for the more orthodox petrol gauge. When the level of fuel in the tank drops to three gallons the warning light starts to flash intermittently and continues with increasing frequency as the level drops even lower. By this means the driver receives adequate warning of fuel shortage, but at night the flashing proved decidedly distracting, and a suggested improvement

would be to arrange the light to start operation at a lower fuel level.

Adequate Weather Protection

The M.G. is a car which will normally be driven in hood-down condition, but the all-weather equipment is nevertheless very adequate. The hood is easy to erect, covers both passengers and luggage and is commendably free from rattle. The sidescreens are a good fit and are rigidly mounted; when not in use they stow away easily in a special locker. A tonneau cover for the small luggage boot is standard equipment, but it might be suggested that, when material becomes in more plentiful

supply, this item could well be extended to cover the seats as well, in order to save erecting the hood when parking the car open in uncertain weather.

With full all-weather equipment erected, the all-out performance of the car is very little affected, and a mean drop of only 0.3 m.p.h. was recorded during the speed test. One down-wind run was, in fact, 1.3 m.p.h. faster than the best run with the screen flat.

A further good feature, while on the subject of weather protection, is the amount of headroom available when the hood is erected.

High averages at night can easily be maintained with the head lights full on, but when it is necessary to dip, the illumination is poor. A satisfactory compromise was, however, found by running with the head lamps dipped and the spotlight set to throw a long beam ahead low down on the road. The limited floor space on the driver's side precludes the fitting of a foot dipping switch. The hand dipper, located behind the horn button, is, however, quite easily operated without removing one's hand from the steering wheel. Limitations of space also make it difficult for a long-legged driver to find a place for the left foot, except over the clutch pedal. It is difficult to see, however, how a more convenient arrangement could be made on the present general layout.

The arrangement of instruments is well thought out, and with the exception of the speedometer all can easily be read by the driver in daytime, but at night the combination of green light on the buff faces and brown figures and hands makes quick reading

ALL-UP. — With hood and sidescreens erected there is ample protection for passengers and luggage, and all-out speed is very little affected.

DBL 606

MORE ROOM. — Greater passenger comfort and slightly more luggage space have been achieved by increasing the body width by 4 ins.

a matter of some difficulty. In passing, it is worthy of note that the speedometer on the car tested was, by present-day standards, unusually accurate, registering only 5 per cent. optimism against a time check.

Interior appointments are of good quality, upholstery is of leather, and the dashboard of polished wood.

To sum up, the TC-type M.G. provides a very adequate means of transport for those who place performance and stability and an ability to go almost anywhere, high on their list of requirements. Mechanical accessibility is above average, and, by present-day standards, the car represents very good value at the all-in price of £527 16s. 8d.

COACHWORK DE LUXE
by
TICKFORD

M·G· TICKFORD D·H· COUPÉ

LONDON SHOWROOMS:
TICKFORD LTD. 6-9 UPPER SAINT MARTIN'S LANE, LONDON W.C.2
WORKS: NEWPORT PAGNELL · BUCKS.

Diagram labels: REVOLUTION COUNTER AND CLOCK · PETROL WARNING LIGHT · SIDE-SCREEN THUMBSCREW · WINDSCREEN WIPER MOTOR · 30 M.P.H. WARNING LIGHT · SLOW RUNNING CONTROL · OIL PRESSURE GAUGE · PANEL LIGHTS · IGNITION WARNING LAMP · STARTER · CHOKE · HORN BUTTON AND DIP SWITCH · IGNITION AND LAMPS · AMMETER · SPEEDOMETER WITH TRIP RECORDER · TRIP RECORDER CONTROL · MAP READING LAMP · BRAKE · CLUTCH · ACCELERATOR · FOG LAMP · INSPECTION LAMP SOCKETS · FOLD FLAT WINDSCREEN THUMBSCREW · GEARBOX FILLER PLUG · FLY-OFF HANDBRAKE

The TC M.G.

MANY years of experience with high-performance cars have resulted in a layout of instruments and controls on the TC M.G. that will satisfy even the most particular sports car enthusiast. Naturally, a stranger to the M.G. would have to familiarize himself (or herself) with the various knobs and dials, but there is nothing complicated about the equipment. There is a reason for everything.

Both the 100 m.p.h. speedometer and the revolution counter are of the Smiths 5-in. type. The former is located on the passenger's side; it has a total mileage indicator at the top and a " trip " reader below. The " trip " reading is returned to zero by a winder below the facia panel. The revolution counter is scaled up to 6,500 r.p.m., and is placed directly in front of the driver; it incorporates an electric clock.

The fully sprung steering wheel is completely " bare," there being no provision for horn button and other switches that are normally found on touring cars. Horn button and dipper switch are combined, and are located centrally on the main facia panel. The dipper switch is of the " flick " pattern; it dips the near-side head lamp and simultaneously extinguishes the other one. Naturally, the dipping arrangement varies according to the lighting requirements of different countries.

Realizing that a sports car must attract attention, the makers have provided a clever 30 m.p.h. warning device. This consists of a green light, which is illuminated when the speedometer is indicating anything between 20 and 30 m.p.h. This warning device is not provided on export models.

The two-way reserve petrol tap, fitted on earlier T-type cars, has given way to a fuel warning light, mounted on the off side of the facia panel. When the fuel in the 13-gallon tank drops to three gallons, this light flashes on and off.

The instruments are illuminated from

behind at night by six panel lights, which give a soft, greenish glow, sufficient to ensure easy reading, but anti-glare at all times. A map-reading panel light is provided; this operates by turning the knob in the lamp body. On export cars, an additional map-reading lamp is substituted for the 30 m.p.h. warning device.

Both the starter and the mixture control are of the pull-out type; the latter has eight progressively r.ch positions, obtained by turning the knob in an anti-clockwise direction. A screw-out slow-running control is also provided. The ignition switch is embodied in the lamp switch, and is wired in circuit with the S.U. electric petrol pump, which remains inoperative so long as the ignition is " off."

The gear lever is of the familiar M.G. remote control type, with the gear positions clearly marked on the top of the knob. A " fly-off " racing pattern hand brake is located to the near side of the gear lever, the ratchet being controlled by a press-in knob on the top of the lever.

The windscreen can be folded flat; it has twin wiper blades driven from a Lucas motor situated on the top of the screen on the passenger's side. Doors are hinged at the rear and contain large pockets for maps, and so on. Thumbscrews are used to secure the celluloid side screens in position; when out of use they are stowed in a special compartment in the rear panel of the body. The hood is folded and rolled behind the luggage compartment, and hood sticks collapse neatly on each side. A tonneau cover fits on press-studs over the luggage compartment.

Although the seats are separate, the squab is in one piece. It can, however, be adjusted for angle, and has several fore-and-aft positions. Seat adjustment is controlled by levers, one on the driver's side, and the other on the passenger's.

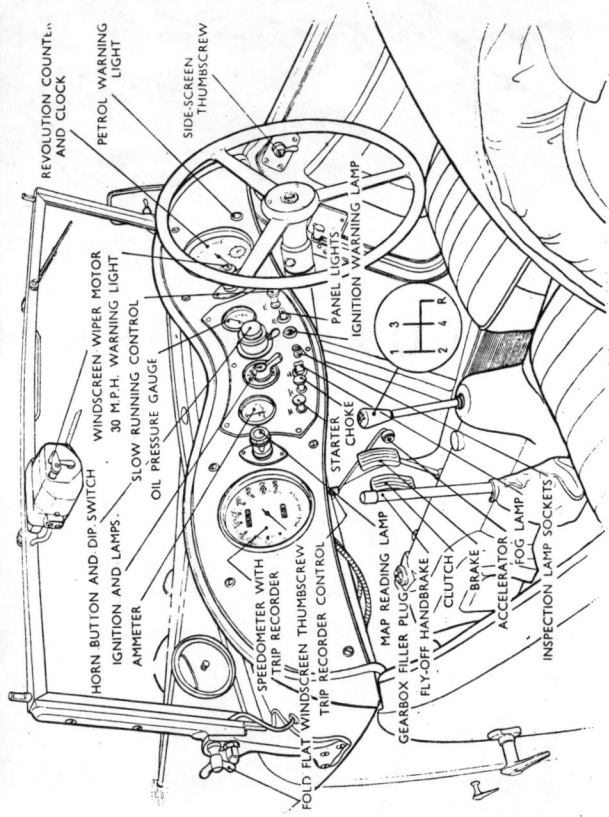

TUNE BOOK

An Official Manual on the Preparation of the M.G. Midget for Competitions

□ □ □ □

an 11 per cent. power increase, and at the extreme end of the scale the high-compression-and-supercharger stage is quoted as giving 80 per cent. more power than normal on an 80/10/10 mixture of methanol, benzole and petrol.

The first stage of tune suggested involves principally the machining of 3/32 in. from the cylinder-head face to obtain 8.6/1 compression ratio, and the putting in of some extra-careful workmanship on such matters as the matching-up of ports, packing pieces being used to correct rocker angles for the amount machined off the cylinder head. Wider tappet clearances and slightly harder plugs are suggested, and the addition of between 50 per cent. and 75 per cent. benzole to the fuel, according to how much full-throttle driving is anticipated.

The second stage of tuning is attained with the maximum permissible amount of metal machined off the cylinder head to give slightly more compression, and

ONE of the most informative books ever published on tuning, a subject all too frequently regarded as a black art, has recently been produced by the M.G. Car Co., Ltd. Dealing exclusively with the 1,250 c.c. engine type XPAG fitted to M.G. Midgets of the TB and TC series, it includes precise facts which eliminate almost all guesswork from the preparing of these cars for participation in speed events, and is a model of service for the keenest type of sports-car buyer.

Sober introductory statements open the tune book, pointing out that "as-delivered" tune represents what the factory regard as the ideal for a car to be used on the road with pump fuel in the tank. The guarantee on a new car expressly excludes any tuning, but it is made plain that factory tests have shown that the power output of the sturdily built M.G. engine can, in fact, be stepped up to meet special requirements without there necessarily being any undue loss of reliability.

Dope for Speed

Tuning largely depends on the availability of fuels better than 70-octane Pool petrol—even untuned, the engine really needs about 74-octane fuel for knock-free running on its $7\frac{1}{4}/1$ compression ratio, or 82-octane optimum performance. For racing use, it will respond properly and reliably to tuning only if fuel anti-knock quality is further stepped up, initially by the admixture of benzole, or to a greater extent by the change-over to methanol fuel flowing through suitably enlarged jets.

Although tuning is a specialist subject, it has in this book been subdivided into five stages, which come near to meeting most requirements, covering three progressive stages of compression-ratio increase and two alternative compression settings for engines running supercharged. The mildest unsupercharged tune stage is suggested as requiring 50/50 petrol-benzole fuel to give

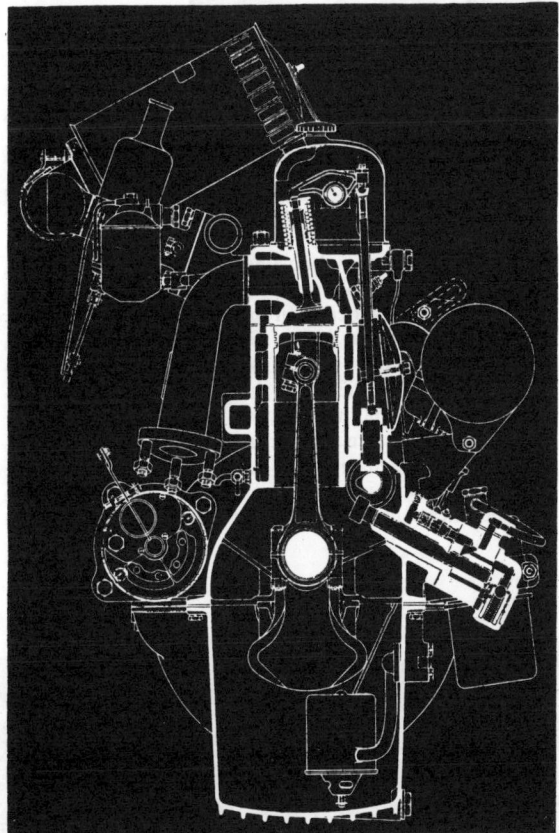

SPEED BASIS.—A section of the standard TC series engine reveals the inclined S.U. carburetters, swept exhaust manifold, and slightly inclined overhead valves.

Tune Book—Contd.

with ports opened out to accommodate larger inlet and exhaust valves controlled by 150-lb. dual springs. At this stage of engine development it becomes worth while to use fuels containing alcohol, for which suitable carburetter needles are quoted, and to fit a second fuel pump to handle the extra consumption necessary with this type of low-calorific-value fuel. Interesting details quoted are that the heavier valve springs, which are of staggered pitch, raise the valve-crash speed to 6,500 r.p.m., well above the normal safe engine-speed limit of 5,700 r.p.m., and also that if speeds are unlikely to fall below 40 m.p.h. an extra 1 b.h.p. may be gained by removal of the fan.

For really high compression, special pistons are needed, those available from the factory giving 12/1 ratio with an unmachined cylinder head and being

SPACE FILLING.—Highest compression ratio is attained with special pistons, shaped to match the combustion chamber and with provision for flame spread from the sparking plug

specially shaped to allow adequate flame spread from the sparking plug. At this third stage of tune, it becomes worth while to change over from 1¼-in. bore to 1½-in. bore S.U. carburetters, and the highest racing-power output quoted for an engine of this compression ratio running on pure methanol fuel is 83 b.h.p. at 6,000 r.p.m., representing a b.m.e.p. of 144 lb. per sq. in.

Any further power boosting depends on use of a supercharger, and the manufacturer's tests have been made with a Shorrock unit belt driven at 1.16 times engine speed, a unit which gives a boost rising from 1½ lb. at 1,000 r.p.m. to 6 lb. at 5,000 r.p.m. Fitted to a standard engine, this supercharger allows 69 b.h.p. to be developed on pump fuel, or 75 b.h.p. on a 50 per cent. alcohol mixture fed through larger jets.

For unrestricted racing, a fifth tuning stage is described, comprising the compression ratio and valve size modifications of stage 2 in conjunction with the

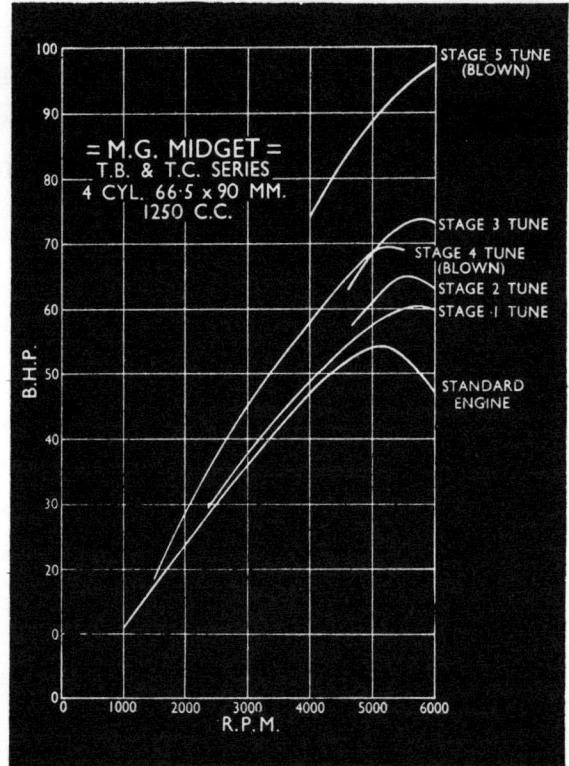

= M.G. MIDGET =
T.B. & T.C. SERIES
4 CYL. 66.5 x 90 MM.
1250 C.C.

STAGE 5 TUNE (BLOWN)
STAGE 3 TUNE
STAGE 4 TUNE (BLOWN)
STAGE 2 TUNE
STAGE 1 TUNE
STANDARD ENGINE

UP AND UP.—Progressive increases in power output at high engine r.p.m., for various degrees of tune with and without supercharging, are indicated on this graph.

supercharger. Using a 50/20/30 mixture of methanol, petrol and benzole, 88 b.h.p. can be obtained, and with fuel containing 80 per cent. alcohol fed through an oversize carburetter the astonishing output of 97½ becomes attainable at 6,000 r.p.m.

A fair number of special parts are needed in tuning an engine and in making a car fit to use extra power, and various such items are catalogued in the tune book. Pistons, valves, valve springs, carburetters, magnetos, sparking plugs, etc., all are listed in their suitable and available forms, plus wheels to accommodate oversize tyres and a high-ratio crown wheel and pinion.

Too frequently the attitude of manufacturers to modification and tuning of their cars by enthusiastic customers has been one of disapproving unhelpfulness. It is pleasant to see that in Abingdon-on-Thames at any rate there is a very much more co-operative outlook, which should bring results by guiding tuners along those lines which thorough bench tests have proved the most promising.

TUNED FOR THE JOB. — An M.G. Midget pursues a Maserati round a corner in a recent American road race.

Dr. Tinsley's M.G. Special

IN 1949, Dr. Henry Tinsley of Belfast was getting from A to B rather fast in his standard TC model M.G., starting the season by winning the 1,300 c.c. standard open car class at the U.A.C.'s Knockagh hillclimb with the record time of 1 min. 16⅔ secs. He followed this up with a class second at the Phœnix Park sprint at Easter, and another at the 500 M.R.C.I.'s "downhill climb" in July. Then came a fifth place at the Ards Airfield races, lapping in 1 min. 35 secs., and a third in his class at the Dublin University M.C. and L.C.C.'s Killakee hillclimb in September. The following week he took a second off his previous Knockagh time, without, however, being placed.

Cheered by these results, the doctor decided to better them by having the M.G. drastically slimmed. Hubert Chambers and Billy Davidson removed the standard body, pruned the front and back leaf springs, fitted heavy wire stabilizers, and increased the castor angle. They retained the standard 4.875 : 1 back end and gearbox, but modified the clutch. As the present clutch is about the fifth which has been installed since the conversion, they are not inclined to talk freely about it! Unsprung weight was cut down somewhat by intensive drilling of the brake shoes and drums, but the normal wire wheels were refitted. The engine then came in for a spot of attention, the flywheel being considerably lightened, and a not-particularly-successful bit of crankshaft-balancing being attempted (actually, this appears to be the opportunity for a pun about forestalling and laystalling, but I can't thing of one). Aerolite pistons were fitted to lightened Morris M.10 connecting rods, while the valve timing was also modified later in the season. A number of cylinder heads were skimmed to give compression ratios ranging from 7 : 1 to 9 : 1, the ports being opened out and M.G. "standard oversize" valves being fitted, with stronger springs. Twin 1½ in. S.U. carburetters were fed from an S.U. pump, and ignition provided by a Lucas vertical magneto off a Scammell truck. Incidentally, the photograph of the engine shows it in its 1,100 c.c. guise, using a Morris block and Riley pistons, while the coil is a temporary fitment.

An interesting feature of this car is the cooling system, incorporating a built-up radiator block and the standard M.G. pump, but no fan. The total capacity of the system is only *one* gallon, yet even with half the radiator blanked off it refuses to boil. The body frame is a mixture of ½ in. steel tubing and box sections, the bulkhead being of 18 gauge Dural, while the body itself is of mere 24 gauge material. Needless to say, it shakes itself to bits in

(Right) View of 1,100 c.c. engine, showing cooling system and large-bore S.U. carburetters.

every long event, after which they cheerfully patch it up again for the next one! An ordinary saloon-type bucket seat is fitted, and the instruments are two in number—an oil gauge and a rev. counter driven off the camshaft. The dry weight is about 12 cwt., which gives excellent acceleration, and maximum r.p.m. is somewhere on the useful side of 6,000.

Thus modified, the M.G. went to the I.M.R.C. hillclimb at Enniskerry on 15th April last year, where she seized solid, but coasted over the line speedily enough to take third place in her class. The following month, at the same club's short Phœnix Park races, the clutch blew up on the first lap, and, possibly as a result of the previous seizure, a couple of cylinder liners came adrift. Undaunted, the boys sloshed jointing compound over the offending liners, and popped them back in again for Craigantlet hillclimb, where Tinsley was third on handicap and third in the 2-litre sports class with a climb of 1 min. 34¼ secs., which was a full 5⅖ secs. better than his 1949 (standard form) time. In the two weeks before the 500 I.M.R.C.'s Cairncastle climb the modification to the valve timing was carried out, and the car won her class with a 1 min. 19 secs. climb. The alterations to the cooling system were made for July, when the M.G. performed well at Ards Airfield, gaining second place in the final. Tinsley brought his lap time down to 1 min. 29 secs., which is worth comparing with his previous year's time over the same course.

R. E. Dorndorf drove the car in the U.A.C.'s Ulster Trophy race over the Dundrod course in August, but on the 8th lap the front carburetter float punctured. Oddly enough, this has occurred no less than eight times, and the Equipe Tinsley confess to being rather puzzled. In September Tinsley was well pleased with the special during practice for the I.M.R.C. race at the Curragh, until one of the alloy rods, which had been fitted for the Trophy, left him in the spectacular manner sometimes adopted by alloy rods. This happened the night before the race, but a search was immediately instituted for another engine. A standard M.G. engine was successfully borrowed and fitted with the requisite parts from the wrecked one, and all-night work brought the car to the starting line. But Fate had not finished with the unfortunate doctor, who left the road whilst lying fourth, with only one lap to go, hitting a bank and smashing a back wheel completely.

The last event of the season was Knockagh again, when the special was run in 1,100 c.c. form, and climbed in 1 min. 13 secs., again without being placed. Now Dr. Tinsley has a new special in the process of being built, with a tubular chassis, the distance between the side-members being the width of the Tinsley posterior. Reversed Vauxhall front suspension is being used, and a Vauxhall engine may also be installed. As he is now the owner of one of the five XK.120 Jaguars in Ulster, it is doubtful if he will retain the M.G. special, although his

only criticism of it concerns a slight lightness in the tail. With improved rear suspension, and a little more development, this car might yet raise a few eyebrows in Ulster motoring circles.

FAREWELL TO A LADY
F.w.d. Derby to be Broken Up

BELOW is the last picture to be taken of the historic front-drive Derby Miller which gained countless records at Montlhéry track during the '30s in the capable hands of Mrs. Gwenda Hawkes. This car held the short track record for many years, only losing it to Raymond Sommer's Type 308 3-litre Alfa Romeo in 1939. Mrs. Hawkes also drove it at Brooklands in 1935. The Derby, with its twin-o.h.c. straight eight engine and centrifugal blower, bears unmistakable signs of its Miller ancestry, and is very typical of its era. As shown, it was still in good form, although the transmission brakes were no longer effective.

The engine will have another lease of life in Formula 1 events next year, for W. R. Baird of Belfast intends placing it in a tubular chassis with 1,100 c.c. Fiat front suspension and de Dion rear, using an E.N.V. 110 gearbox and two-stage supercharging.

MG - BUGATTI

Unusually neat installation of a Type 40 Bugatti engine in an MG TC

Article and Photos by Jack Campbell

Should you, in your meanderings, chance to come upon an MG TC that sounds as if its vitals would at any moment be strewn about the landscape, be not fooled, for beneath the bonnet there beats a true and noble soul, the spirit of "Le Patron." Yes, an erstwhile docile little MG has, thru the wizardry of Eugene Marsh of Long Beach, California, been transformed into a crotchety and screaming prodigy, by the installation of a Type 40 Bugatti engine.

Now for those of us that are not up on our Bug background, the type 40 was built in the years between 1926 and 1930. It has a displacement of 1496 cc, a bore of 69 mm, and a stroke of 100 mm. The engine is a four cylinder, with an overhead camshaft that operates three valves per cylinder two intake and one exhaust. The crankshaft has five main bearings of the poured babbit type, as are the connecting rod bearings. The head is irremovable cast in one piece with the block—Bugatti style. The Works rating, which was unusually conservative, set the horsepower at 50, with a compression ratio of 5.7 : 1. At present, this type 40 is packing 8.5 : 1.

The overhead cam is driven by a shaft located at the front of the block. This same drive operates the magneto and a twelve-volt aircraft generator.

Mounted on the battery box is a two-gallon reservoir which takes the place of part of the original oil sump that was chopped off by another owner to make the engine fit into a midget race car chassis. (Le Patron just red lined his rev counter.*) A modified oil pump surges the lubricant thru the lines at a mere 150 lbs per sq in. at idling speed . . . . this feature is going to be altered to reduce the pressure.

Getting the Bugatti engine into the MG chassis was no small task. There was an extra half inch in the length of the engine that complicated matters. But with prudent cutting the engine was recessed to clear the front cross-member. The only changes in the chassis concerned the steering column, which had to be moved forward and down. The ball joint on the left front wheel was inverted so the drag link could pass below the sump and front spring.

As far as the weight and positioning of the engine is concerned, there is very little change from original. The engine weighs almost exactly the same as the MG, and Marsh used the stock MG clutch, making an adapter to match the bell housing. The handling characteristics have not been altered, save in the direction of added urge.

Not enough can be said concerning the masterful way in which Mr. Marsh went about this job. The workmanship is the best, and Ettore Bugatti can rest a little more in peace, knowing that one more of his little jewels has come into the hands of one who knows.

*Competition slang for "flipping his lid."—Ed.

TEMPERATURE RECORDING ON A TC M.G.

By

Philip H. Smith, A.M.I.Mech.E.

IT will be generally agreed that the provision of water and oil temperature recorders is a desirable feature of a completely equipped sports-car. Many cars, however, are not so fitted as standard, and although suitable thermometers of a type to match the instruments normally on the panel are usually obtainable as an extra, their cost is liable to put them in the "luxury" class. An alternative is to utilize the Government surplus instruments which are widely advertised and easily obtainable at bargain prices. These thermometers, whilst of extremely high grade manufacture, and absolutely reliable, have three general objections to their use, in that (a) their dials do not match the other instruments, (b) there is a colossal length of capillary tubing to accommodate, and (c) the oil sump must be removed in order to drill and tap a hole for fitting the element (this applies, of course, irrespective of the type of instrument, unless the sump is already provided with a blanked-off union).

These objections can be overcome by a little preliminary designing, and this description of fitting the instruments to a TC type M.G. will be useful to readers with similar ideas.

After considering various alternatives, it was decided that a good-looking and practical job could be made of the installation by mounting the two thermometer dials on an auxiliary panel, in a position where their non-matching finish would not be so apparent. As regards the excessive length of capillary tubing, investigation showed that there was sufficient spare room in the battery compartment under the bonnet. Finally, manufacture

The auxiliary panel, containing the clock-dials for the water and oil temperature gauges.

of a special adaptor taking the place of the sump drain-plug enabled the thermometer element to be fitted without removal of the sump.

Mounting the Dials

The type of instrument used was the R.A.F. pattern made by Coley, and reading 40 to 140 deg. C. It is agreed that the top reading is unnecessarily high, but previous experience with a dial having a "max" of 100 deg. C. which after about half a minute under boiling conditions resolutely refused to zero again, influenced the choice so far as water temperature was concerned. The "bakelite" cases of these instruments are provided with square fixing flanges, drilled for four screws with their centres on a $2\frac{1}{2}$ in. square. Only three of the fixing holes were used in each case, the bottom corner, which came on the outside of the panel, being removed with a hacksaw to fit neatly into the outline of the panel, as shown in the photograph.

The panel is made of $\frac{1}{8}$ in. duralumin, the two thermometers being inserted into their apertures from the rear after unscrewing the rims securing the glasses. Three 2 BA nuts and bolts secure each

instrument, the two top outer ones being specially long so as to pass through mild steel angle brackets which attach the panel to the batten running along the bottom of the M.G. facia panel. The position chosen for the mounting, just to the right of the steering column, and allowing sufficient room to get at the clock-winder, does not interfere with access to the car, whilst at the same time the dials are unobtrusive and easily read. Illumination is not considered necessary, as the figures are luminous.

Layout of the Tubing

The capillary tubing is fairly flexible, and will stand quite a lot of bending, but care must be taken not to kink it. In particular, extreme caution is necessary at the point where the two ends of the tubing join the units. If the tubing is broken, the instrument is useless.

The tubing is supplied wrapped on a reel, and for a start, it should be uncoiled, using as many hands as are available to control it, and plenty of room. From the panel, the tubing is run parallel to the steering column, through the rubber boot where the column passes through the bulkhead, across the

(Above) The special adaptor for the sump drain plug.

(Left) Capillary tubing junction with the battery box.

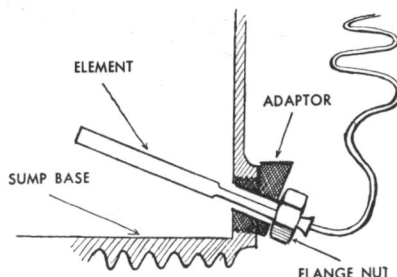

The sump drain-plug adaptor permits the element to lie in an adequate depth of oil.

Temperature Recording—*cont.*

front of the battery box and thus to the nearside of the car. It should be fed through the bulkhead before any attempt is made at neat installation. Once through, the tubing can be secured to the steering column by a couple of jubilee clips. If it is run along the top of the column, that is, at 12 o'clock on the column's diameter, it will be quite unobtrusive. To allow the elements to pass through the bulkhead, it will be necessary to slacken off the plate securing the rubber boot. A few anti-vibration coils can be made in the tubing between the steering column clip and the panel, and the latter finally secured in place.

Examination of the interior of the battery compartment will show a space between the end of the box and the battery sufficiently large to accommodate the excess tubing without disturbing the battery. If necessary, the latter can be moved as far as possible to the offside and the holding rods repositioned, to allow more space at the nearside. The excess tubing is then made into two coils—one for each instrument —of as large a diameter as can be accommodated in the space. Copper wire wrapped at intervals around the coils will help to hold them neatly. Before installing them between the battery and the box-end, make sure that sufficient free length has been left, allowing for anti-vibration coiling, to reach the radiator header tank and sump drain plug. It will be clear that there are four tubes, two in and two out,

emerging from the battery box, and these should be wrapped tightly with copper wire for an inch or so to add strength and to hold all four neatly together. A rubber or felt sleeve over the wire wrapping will prevent chafing when the box lid is in place. The latter will have to be slotted to allow the tubing to pass, and a neat job can be made by suitably shaping the slot and providing it with a hooded outlet, as shown.

Fitting the Elements

The radiator header tank is provided with a blanked-off union threaded $\frac{3}{8}$ in. B.S.P. After removing the blanking nut, it will be found that the element, which has the correct size of gland nut, can readily be fitted. The tubing is run along the tie-rod between bulkhead and radiator, and secured to the tie-rod by three small clips. About six large-diameter coils are made in the

tubing between the header tank and the top of the tie-rod.

As regards the sump fitting, this makes use of the drain plug orifice. It is, however, undesirable for the element to lie parallel with, and close to, the bottom of the sump, as would happen if the drain plug was merely drilled for a $\frac{3}{8}$ in. union. In this position, there would be no depth of oil under the element, which would be in the coolest position, and thus give a false reading. An adaptor is, therefore, manufactured which, whilst threaded to fit the drain-plug hole, is itself drilled at as acute an upward angle as possible so that when in position, the element lies at an inclined upward angle in the oil, its outer end being some two inches from the sump bottom. The brass adaptor shown allows an angle of 15 deg. from the horizontal, the threads thereon being $\frac{1}{2}$ in. B.S.P. to fit the sump thread, and, of course, $\frac{3}{8}$ in. B.S.P. for the element gland nut.

Draining the Sump

It will be noted that the adaptor has a small drain plug fitted in the side. This is threaded No. 0 B.A., and although admittedly small, will actually drain the sump in a night. It is, therefore, used for a normal routine oil-change, while at longer intervals, as when the filter element is renewed, the sump adaptor is removed bodily to provide a "major" drainage.

In view of the flexibly-mounted engine, plenty of spiralling should be left on the tubing leading to the sump, but so long as this is done, there will be no danger of fracture.

When all the tubing has been finally coupled up, it should be clipped in position along the bulkhead, as convenient. Unwanted kinks should be straightened out and all bends made as symmetrical as possible. Needless to say, insulation tape as a fixing medium is barred.

Apart from its merits as a practical and neat job at low cost, the installation has the advantage that it can be removed and transferred to the new car (when it arrives) without leaving unsightly holes in the instrument panel, or similar undesirable evidence.

USED CARS ON THE ROAD

THE AUTOCAR, JUNE 20, 1952

No. 35: 1947 M.G. TC Two-seater

Price, new: £412 10s plus £115 6s 8d purchase tax.	Acceleration from rest through gears to 30 m.p.h., 6.4 sec.	Fuel consumption range: 28-33 m.p.g.	Speedometer reading: 17,500.
Secondhand : £575	To 50 m.p.h., 15.9 sec. To 60 m.p.h., 24.0 sec. 20-40 m.p.h. (top gear), 14.6 sec. 30-50 m.p.h. (top gear), 13.6 sec.	Oil consumption 1,500 m.p.g. approximately.	Car first registered June, 1947.

AS has already been recorded in *The Autocar*, the outstanding total of 90 per cent of the present M.G. two-seater production is not only being exported but also sold for dollars. The popularity of the Midget has remained very high since the model was introduced, many years before the war, and it is a significant indication of the recent drop in second-hand car prices that the car tested was offered at a price less than £50 in excess of its original cost, and considerably under the present total for the current TD. Within this price was also included a number of extras. The place in the sun which the M.G. has earned by a process of honest and logical development is such that there is now scarcely any other car which can provide this type of motoring at similar initial and running costs.

The appearance of the car tested, which was supplied for the purpose by Dicks Car Sales, Ltd., 385-401, High Road, Kilburn, London, N.W.6, was quite unusually good. The black cellulose had a deep lustre and all the chromium plating was without flaw. The interior was clean, fairly new loose covers having been neatly fitted to the seat cushions and one-piece squab. A new tonneau cover was included in the weather equipment and the hood and side screens, although showing their age, were still sound.

In general the performance was excellent. Starting was always immediate and the engine quickly pulled well without use of the mixture control. The engine was flexible on all the gears but use of the gear box provided that liveliness in performance which enabled the car to dart through traffic or eat up distance on busy main roads with its traditional willingness. Only the tickover seemed a little unhappy, having a tendency either to run rather fast or, when hurriedly adjusted, to stall. Everything about the car suggested thorough servicing, the brakes being a good example. They were adjusted to a nicety and were really powerful and smooth under all conditions. The fly-off type of hand brake lever also worked well. The steering, however, was a little over-tight, being stiff for manœuvring in confined spaces, and a little jerky on the road. This tendency gradually diminished during the test. This TC model has, of course, the traditional beam axle at the front, giving a firm ride but not at all an uncomfortable one for the type of person who enjoys sports motoring.

The tyres, although worn, had a fairly useful amount of tread remaining. The comprehensive instruments worked well, only the speedometer being a little jerky. Accessories included an unusually good tool kit, a massive exterior mirror, a radiator mascot, Trafficators with combined unit of tell-tale light and control switch, and a luggage grid above the petrol tank. The car was healthy, efficient and thoroughly smart.

CONTINUED FROM PAGE 66

6. By removing the guards, lights, bonnet, generator, windscreen, starter motor, fuel tank, one floorboard, and having alloy door skins and lighter battery, a TC weighs 11 cwt. Road tested in 1946, a TC did 78 mph.

7. Some drivers fit TD headlights. They reckon they work better, are smaller and lighter.

8. Many, many owners have had a wheel fall off whilst driving. The hubs and splines don't cost a lot to be built-up.

9. Oil: Expert mechanics say that a TC sump is one of the trickiest to put back. Boy, do they leak! The back hub gaskets are made of paper. They leak too. It pays to put two gaskets instead of one.

10. On the front axle above the spring, there are special plates to give a flat surface when the suspension bottoms. But the axle already has a flat surface so really, they're useless. Why not throw them away?

But don't throw the TC away. They'll never replace the motor car, but there's nothing quite like them.   #

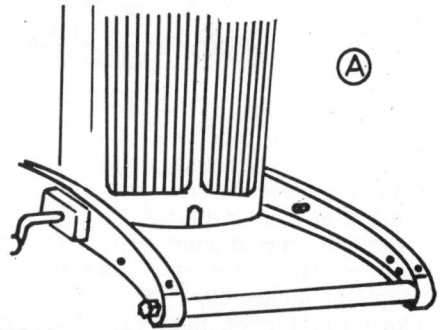

Modifying the front chassis ends: Pic A shows standard front end, and pic B the full boxing, square tube cross members and dampers.

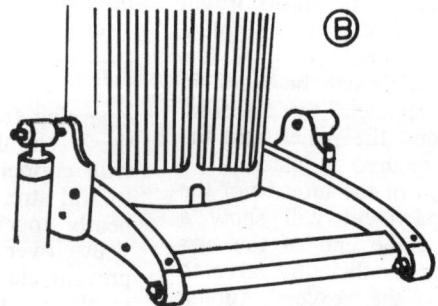

The full treatment: A South Australian TC at Mallala with 15in. wheels, lowered radiator, and high-mounted telescopic dampers.

this may have happened to you

By Bill Pollack

the GOOD PENNY

Is THERE such a thing as reincarnation for automobiles? This thought occurred to me one day as I was wondering what ever happened to my first sports car. To those few people who have owned a genuine sports car there is a fondness for the first one that is never exceeded by that for later models no matter how outstanding. Perhaps this is because that first car was your teacher. It was the one that opened your eyes to real driving pleasure; it broke you of those habits acquired from driving the larger variety of transportation; it had patience with your fumbling gear changes; it forgave you for those ambitious corners and, most of all, it was the sheer pleasure that is found only in the "first."

The other day we ran into an old friend who had a look of pure joy on his face, and a glance behind the radiance exposed a TC-MG with the original dark green paint; it had cut-down, 16-in. wire wheels and the engine still boasted a Nordec supercharger. "It's not the same," "where," "how," "who," were some of our strangled questions. In the answer lies a story.

Early in 1947 a freighter docked in Los Angeles. Among the manifold items in its cargo was a shipment of twenty-five MGs, all painted red but one. This singular car had that beautiful dark green found on the early TCs. The original owner found the car completely impractical, hard on his spine and it did not get the thirty miles to the gallon that the salesman claimed. This man had the car a few short weeks and then sold it to our friend. The new owner looked at this little bit of steel and green paint as the eighth wonder of the world and thereupon began a program of care. The car was polished every week end; if the surface was dirty, then only cold water, a sponge and chamois were used. For this man, lubrication was not a matter to be trusted to individuals who might in a rare instance miss a fitting. A grease gun was purchased, including a complete line of lubricants from gear grease to rubber lube. Our enthusiast went so far as to buy medicinal glycerine and graphite which he combined with a light solvent to make his own rubber lube. The leather was cared for in a manner which would amaze modern skin specialists. The engine always spotless, from time to time the power plant would be completely gunked, scraped and re-painted in the original color. To preserve the wood components of the car, every exposed piece of wood was given coats of varnish with special attention to the underneath part of the floor boards. The body bolts received their share of time and the chrome was cleaned a la MG Manual with water and cloth only. All of this care was dutifully recorded on the Service Data sheet in the manual.

In addition to the care, the car gradually underwent many modifications. The wheels were removed and cut down in size for added strength and beauty. The engine came into its own with a brand new Nordec supercharger. Then came a long line of accessories but always only those which were in good taste. Finally, the car was driven. Hardly a day passed that the owner did not have a "go" with someone. This little car was not placed on a shelf or kept in the garage for Sunday driving only. It was used every day for work and practically every night for pleasure. A run over the canyon, which amounts to twenty miles in this part of the country, for a quart of ice cream was not an uncommon occurrence. The very first organized road race in the West found this car entered. It ran, finishing second in its heat, and from then on every event found this dark green TC on hand until the growth of "specials" made it impractical to compete.

The MG was finally traded for a Lagonda and then came a Riley followed by an assorted variety including more MGs and ending with a Porsche coupe. The Porsche achieved a high status with this owner for it marked his return to competition. The little coupe became a familiar sight in western competition. News arrived of a new super Porsche with more horsepower. Plans were laid, the first being the sale of the successfully raced Porsche. The car was placed on consignment but no sale. Too many people were aware of the fact that the car had been used in competition. This fact detracted from, rather than added to, its value. The average buyer usually fails to realize that a car must be kept in excellent condition to compete successfully. An advertisement was placed with one of the daily papers and early the following morning a prospect arrived, a car was parked at the curb, a trade-in perhaps, an MG-TC, supercharged, 16-in. wheels, original top, dark green paint still perfect, accessories all there. Same car? Of course. Was there a sale? Perhaps the term "sale" does not adequately describe this transaction which was marked by a lack of conversation and grins on the parties' faces. The prospect? Oh yes, he had seen the MG race and remembered the owner.

epilogue

How many people actually owned this car is something we will never know, but the intriguing question is why the car was kept so perfectly intact. Perhaps a little bit of the thrill and wonder experienced by our friend remained with the car. Surely the work, love and patience that went into it were evident to even the most calloused individual. Each owner must have felt a duty to preserve that which others preserved before him. Reincarnation? A bit far-fetched but a good penny? Maybe . . .

C'est Formidable!

11,000 miles in an MG-TC through Europe's most beautiful countrysides

Text and Photos by Miles C. Wambaugh

I wanted to throw my arms over my head and hide the first time an omnibus came down the road, intent on passing to my right. My wife-navigator kept chanting "keep left, keep left" and, with vibrating nerves and clenched jaw, I managed to "keep left" during those first few terrifying hours in English traffic.

Each crossroad's turn required a complete stop and deliberation lest a new course find me on the right (wrong) side of the road face to face with an angry truck. It was a full day before I gathered enough pluck to pull right and pass a car for the first time. But taking traffic circles to the left produced a dizziness that I never did overcome.

It was only a month before garnering these first impressions of driving a TC on its native English soil that we decided to take a grand tour of Europe by car—or rather by MG, which is not precisely the same barrel of mackerel.

The baggage problem demanded some sort of luggage rack. Since the stock MG racks that pile bags on top of the spare wheel have never appealed to me, an invention was clearly in order. We removed the spare wheel carrier, constructed a semi-permanent rack on the frame ends, and bolted the spare wheel to the back of the rack, Continental style. The bottom portion of the new compartment had a locked steel box fitted for spare parts. In the well behind the seats, by judicious choice of bags and plenty of leverage, we fitted two suitcases, a typewriter and Speed Graphic, a two-quart can of oil and collapsible canvas bucket, a satchel of travel books and one of film, a bag of personal and automobile documents, various sweaters and gloves and two overcoats. Total weight of luggage: 250 pounds. With no passengers the rear springs drooped discouragingly, and we never dared look at them with two passengers aboard. It

was actually a pleasant surprise when the front wheels remained on the ground.

We landed late Saturday evening at Southampton, where everything ran amuck. Our British registration, international and British driver's licenses and insurance policies—for which we had paid before leaving home—were supposed to be waiting for us with the Royal Automobile Club port agent. No such luck! Nobody had heard of us. I raged futilely up and down the pier while my wife calmly inquired after a hotel and learned to operate the English coin telephone with borrowed pennies. The next morning we were treated to an example of how very helpful the British can be. On the strength of my cash receipt from the American Automobile Association for "foreign travel documents," the Royal Automobile Club agent issued all necessary registrations, licenses, and a temporary insurance cover.

"Cheerio," he said, and we drove off with British plates and a "GB" (for Great Britain) securely attached. My regular papers arrived that week from New York.

We headed west to Cornwall, and soon discovered the well-paved English road is a narrow twisting thing with high hedges on each side. The quick steering ratio is a must for, when a big, doubledecker bus comes trundling around a corner, you jump for the ditch and fast. In many sections the roads climb and descend almost on the perpendicular, and it is here the four-speed transmission comes to the rescue of an overburdened, peanut-sized engine. The first time we filled up with gas we realized why the Englishman prefers a watch-charm engine . . . gasoline cost and the tax on horsepower!

Driving in England has many good points. The English are probably the world's most courteous drivers, and we

*Gem-like Switzerland with its care-
fully maintained roads offers much
to delight the sports car driver*

soon found ourselves saying "After you, Cedric," at inter-
sections. Every British driver indicates his intended direction by popping out little lighted traffic arms which, oddly
enough, never seem to strike pedestrians in the eye. Bus
and truck drivers are invariably helpful about waving a
following car on, often into the face of approaching traffic.
The grade-crossing question they handle neatly by fencing
off the trains with a gate on the track which is not
opened unless a train is due.

Entering London at the height of the outbound traffic
rush is a frightening experience. We followed a single
lane into town with flocks of busses, trucks, autos and
bicycles springing at us from all directions, while motor-
cycles weaved in and out of the mass like minnows running from a school of fish.

For a lover of modern and vintage sports cars there is
no country to compare with England. In France we saw
but few Bugattis; Italy showed us only one 2.3 Alfa "Monza":
and we searched Germany in vain for an SSK Mercedes.
But in England we were constantly exposed to the best
examples of the Continental species as well as the numerous
British *marques* that make the enthusiast's mouth water.

On June 3, we shed our bowlers, tweeds and canes and
stepped aboard the Dover ferry. By this time we were
accustomed to seeing the MG suspended in mid-air at the
end of a cable, like an outsized yo-yo; but after it was
lowered into the hold, we were dismayed to see our car
and a Morgan "Plus 4" dicing around below decks with
stevedores behind the wheels. Less than two hours later I
was roaring up the streets of Calais on the left-hand side,
while gendarmes gesticulated with their white wands and
horns were blown as only French drivers can blow them.

The Frenchman is a virtuoso on his horn, and little wonder.
He practices constantly. He lives in fear that something
will sneak up on him before he gets a chance to blow his
horn at it. The French people, particularly the young fry,
seemed to be fired to enthusiasm by the sight of our noisy
little yellow monster on wire wheels. Whenever we stopped,
we were surrounded by children (who pressed in close
but didn't touch!) asking whether our mount was a *voiture
de course* (racing car).

Such is the frequency of kilometer posts beside the roads
in France that one suspects some sort of a national fetish.
Surveying must at least be the national outdoor hobby.
As the posts flashed by we would read, "St. Etienne, 3.5"
. . . "St. Etienne, 3.2" . . . until St. Etienne had passed
without our noticing anything unusual.

Imagination shown by the French in methods of transportation kept us constantly agog. A family of two adults
and four children on a motorcycle and sidecar became com-
monplace. But a man on a bicycle towing his wife behind
in a home-made trailer was more unusual. Later we saw an
elderly man on a motor-scooter, carrying a large dog in a
knapsack on his back like a papoose. A farmer in an ancient
Citroen whizzed by with his dog poised and braced on the
hood, a live radiator ornament. We never did grow used to
the sight of a nun on a motorbike, or a priest, complete
with goggles, streaking for salvation on a motorcycle with
his cassock ballooning out behind.

The French bicycle race is something to steer clear of
if you are in a hurry. We ran afoul of the same race six
times on a weekend in Brittany and each time were held
up at least two hours waiting for the profusely-perspiring,
madly-pedalling horde of competitors.   CONTINUED

*Our MG visited fishing villages of Brittany, the Jungfrau in
Switzerland and the lush woods of Germany's Black Forest,
where elaborately carved wooden signposts direct the tourist*

C'est Formidable

This small main road into a Corn-wall fishing village proved the value of maneuverable English sports cars

In France and the rest of Europe we usually put up at hotels in the smaller towns, chiefly because they are cheaper and rooms are more readily available without reservations. For the entire four months, except for London and Paris, we traveled without reservations and were never stuck. Almost every rural French hotel provides a closed court or a garage for the car, and from time to time the MG was bedded down with busses, crates of vegetables, or heaps of delicately-scented fish nets. On one occasion in Brittany, in a barn attached to the rear of the hotel, we left our car nervously watched by three cows, a plow horse, a trotting racer and a dozen hens, one of which was seated grumpily behind the wheel when we appeared.

Eventually, of course, our path led to Paris and the world's most famous traffic. We developed, and present here, a little set of rules for Paris driving, which saw us through with no crashes and only average nervous strain. *Rule 1:* Never stop or even hesitate, for then not only are you lost but the entire system breaks down and the chain reaction will cause accidents all over Paris. *Rule 2:* Adhere strictly to the Paris rule of right of way which means, ignore the guy on your left and give the fellow on your right no more than an even break. *Rule 3:* Always keep your car in first or second gear, for quick acceleration is often imperative. *Rule 4:* To get out of a tight spot, give it the gas, never the brake. *Rule 5:* When cornered, select a bus or truck for use as a blocking back. *Rule 6:* If, by bad luck and mismanagement, you find yourself near the center of a traffic circle, don't try to do it all at once but continue around, improving your position by one lane on each lap, until you reach the perimeter. Another solution calls for increased acceleration until centrifugal force throws you out of the circle; but this method is best left to the Parisians. *Rule 7:* Convince your wife to sit under the dashboard, where she will be safer and less likely to break out screaming . . .

Over all this pandemonium and amateur grand prix the Paris gendarme watches with momentous calm and detachment. He seldom interferes because the system seems to take care of itself. From 125 miles of the fiercest Paris driving we emerged unscathed. Only three taxis tried deliberately to run us down, and they missed.

When we crossed the border into Italy we knew immediately that there had been a change. The Frenchman drives his car hard for a stretch, and then nurses it a while. The Italian drives flat out all the time, whether it is a sports car, family sedan, *topolino* Fiat, motorcycle or scooter. He thinks he is letting the vehicle down if he isn't getting the most out of it at any given moment. Among the Italian people is found a true appreciation of the uninhibited exhaust, since a straight pipe makes music to their ears, and the shorter the pipe the better. I saw a white-garbed policeman in Padua cock his head attentively and murmur *buono* at the sound of an ear-shattering exhaust that would send a New York or Los Angeles rookie policeman screaming for the riot squad.

The MG was handsomely greeted throughout Italy with waves, shouts and similar evidences of admiration. Our greatest response, however, came in the medium-sized town of Carrara. It was five in the afternoon, with many people in the streets, and we were moving slowly through town. A man in front of a cafe waved cheerily and disappeared inside, to emerge a few seconds later with twenty friends, all of whom set up a shout. Down the road another group rushed out of a cafe and lined the sidewalk to see what all the commotion was. I gunned the engine a bit, and drew an appreciative babble from the crowd. Smiles flashed back and forth between us and hands were raised in the air. The policeman on the corner waved us through, imitating the checkered flag routine, and we roared out of town wound up to 4500 rpm in second gear while the cheers and hubbub died behind us. My noisy muffler and 2½-in. tailpipe had done their job well. Ascari himself couldn't have received a warmer reception in his home town.

It was in this area, near Rapallo, that we experienced our accident of the trip. It happened while we were parked by the side of a mountain road gazing at the Ligurian Sea some 1000 feet below. As I started the car to leave, my navigator asked, "Are you sure there isn't a stone marker post in front of you?" "Don't be silly," I replied, the way husbands do, and took off in a burst of speed that ended abruptly six feet away. The front bumper and supports were grievously damaged but not, I think, as badly as my dignity.

For each new country entered I made a sincere attempt to learn the words for "go

to hell," but this proved unnecessary. In Italy the words for road directions were also unnecessary, since the Italians by reputation and in fact are the world's greatest users of sign language. At the faintest suggestion of where you want to go, he will take you down the main street, through side streets, over bridges and under tunnels, past churches and olive groves, all with his hands. This is accompanied by a running commentary in Italian which can be disregarded, so perfect is his language of gestures. Time and again we followed such directions with complete success.

Coming into Florence we crossed the huge square in front of the hotel where middle-aged women on bicycles began to draw alongside bearing news of "cheap hotel room, very nice, very beautiful view." It was impossible for us to cross that square without feeling like something out of that famous Italian film *The Bicycle Thief.*

From Florence to Forli and Ravenna we had our most breathtaking drive in Italy, crossing the Appenines that form the mountainous backbone of the country. An hour out of Florence we were high in the hills, looking down on olive groves and farms dwarfed by our great height. The outer circle of our view was rimmed with billowing white clouds, while the warm Italian wind blew softly.

We entered Switzerland by the 6000-foot Simplon Pass over a road of packed gravel. It is impossible to drive anywhere in Switzerland without feeling like an Alpine Rally entrant, pouring on a little extra coal on the mountain switchbacks and winding her up a little further through the gears. It may have been this surge of sporting blood that brought on our first mechanical difficulty: the MG, rebelling at last against its excessive load, came down with a bad case of broken spokes in the rear wheels. We coasted gingerly into an MG agency, providentially located at Spiez, where new spokes were fitted.

It was in Switzerland, also, that we first became aware of the tremendous bond that exists between drivers of TC MGs abroad. Never did a TC pass us in either direction without a wild wave from the driver. How the Swiss TC drivers manage to wave the right hand vigorously, blow the horn, and maintain control of the car is a mystery to me. I tried it and ran out of hands.

Then into Germany to do battle with diesel trailer trucks and the *autobahns.* The German truck, moping along and blackening the countryside, will haul as many as three huge trailers. We looked like coal miners on a Sunday outing after gazing into a few of these trucks' tailpipes, which are just high enough to blow in over the door of the MG.

The happier side of motor travel in Germany was presented by the *autobahns.* The surfaces are growing old now, with many patched areas, while the job of replacing war-destroyed bridges is only partially completed. Curves and grades are so beautifully engineered that 60 mph is a comfortable loafing average for touring. From the lush woods of the Black Forest to the rolling Eifel hills, from the legendary Rhine country to the sweeping yellow fields near Würzburg and the breathtaking Bavarian mountains, we found that Germany contains the most lovely countryside in Europe.

Quitting Germany in a cloud of Rhine wine and song, we entered Holland, a land of good roads, good food, good scenery and good people, almost all of whom speak English. The surface that a Dutch roadbuilder can produce from bricks is found in the United States only on billiard tables. My perceptive navigator pointed out that Holland is a country that looks just as everyone always imagined it. The windmills still gather in the breezes with their long arms, and canal barges move gently down the main streets of the cities, while wooden-shoed nationals eat raw herring at street corner stands. The

whole country is so crisscrossed with canals that we stopped more often for drawbridges than for traffic lights. Here the bicyclist is king, and cars are allowed on the public highways only as an afterthought. In most places the motorist is now being liberated by separate pavements for bicyclists, who have a habit of riding hand in hand, eight or ten abreast.

In Belgium there are two grades of road: pretty good and just plain terrible. The latter is of a peculiar cobblestone construction called *pave*, a secret known only to the Belgians, with whom I fervently hope it stays. After 75 miles of jolting over the cobbles at idling speed, my navigator threatened to start tearing up the stones and heaving them at the heads of the passing Walloons. Fortunately, by this time she was too stiff in the joints to get out of the car, and a sticky incident was avoided.

Bucking a heavy wind and deluges of mixed rain and sea water, we worked our way down the coast to Dunkerque, where we drove the car aboard a ferry.

September chilly winds were striking the bones of Englishmen when we arrived, and tops on open cars had been universally raised against the cold. More than one blue-nosed Briton, sealed in his Morris saloon, mentally sized us up for the loony bin as we breezed by with top down and scarves flying. On our part this wasn't totally a desire for fresh air and rosy cheeks. We wanted to see the sights and knew from experience that driving an MG with top and side curtains raised is like traveling down an endless, canvas-lined tunnel.

Farther north in the Scottish highlands the weather was even brisker, but the true Scot considers it a point of honor to keep his automobile top down, his house windows open and his fire dead in the grate. I'm sure the Picts and Scots who swarmed over the old Roman Wall didn't paint themselves blue. They were naturally that way from the cold. In the highlands we found the

CONTINUED ON PAGE 23

Dutch drawbridges are a deterrent to rapid travel in the lowlands, while French river banks are excellent for camping. In Scotland a simple ferry was able to transport two cars

INTERESTING COMPETITION CARS

A REVIEW of the noteworthy sports-racing cars produced in the post-war years would not be complete without reference to the Lester-M.G. The outcome of considerable experience with M.G. cars, the first Lester-M.G. was produced by Harry Lester in 1949 and ran in prototype form at one or two sprint meetings during the summer of that year. The engine used was a TC M.G. bored out to 1,467 c.c. and this was, in fact, the first engine to be so treated. During the following year J. C. C. Mayers drove one of these cars for Harry Lester and in 1951 Pat Griffith joined the works team. By now the cars had become faster and faster and appeared more and more frequently in race results. In 1952, the Monkey Stable was formed, three Lester-M.G.s being driven by Mayers, Griffith and J. Ruddock; and at the end of the year the team cars were sold, thus closing the first chapter of the Lester-M.G. story. The cars had been successful and, looking back on race reports of that period, one finds few that do not draw attention to their performance.

New Design

For 1953 the team transferred its allegiance to Kieft, and at the end of the year it was decided to commission Harry Lester to design a completely new car for the Monkey Stable. Work went on through the following year, until the end of 1954, that is, on the prototype car. At this stage the Stable took over the construction and one almost completed chassis, with the intention of building two 1½-litre cars, using TC M.G. engines with an even greater power output, and two 1,100 c.c. versions with Coventry Climax engines. Team drivers this season for these cars will be J. C. C. Mayers, M. Keen, T. Line and M. Llewellyn.

The basis of the new car is a straightforward welded tubular steel chassis frame of Lester design. The side members are of 3½in diameter 16 s.w.g. tubing, swept slightly upwards towards the front from a position about halfway along the car. The front cross member is similar to the standard M.G. pressing of deep box-section form. It houses the top abutments for the suspension springs and is suitably constructed to carry the anchorage points of the wishbones. An intermediate cross member, the same diameter as the main side members, is provided immediately aft of the gear box and carries the rear mounting points of the power unit.

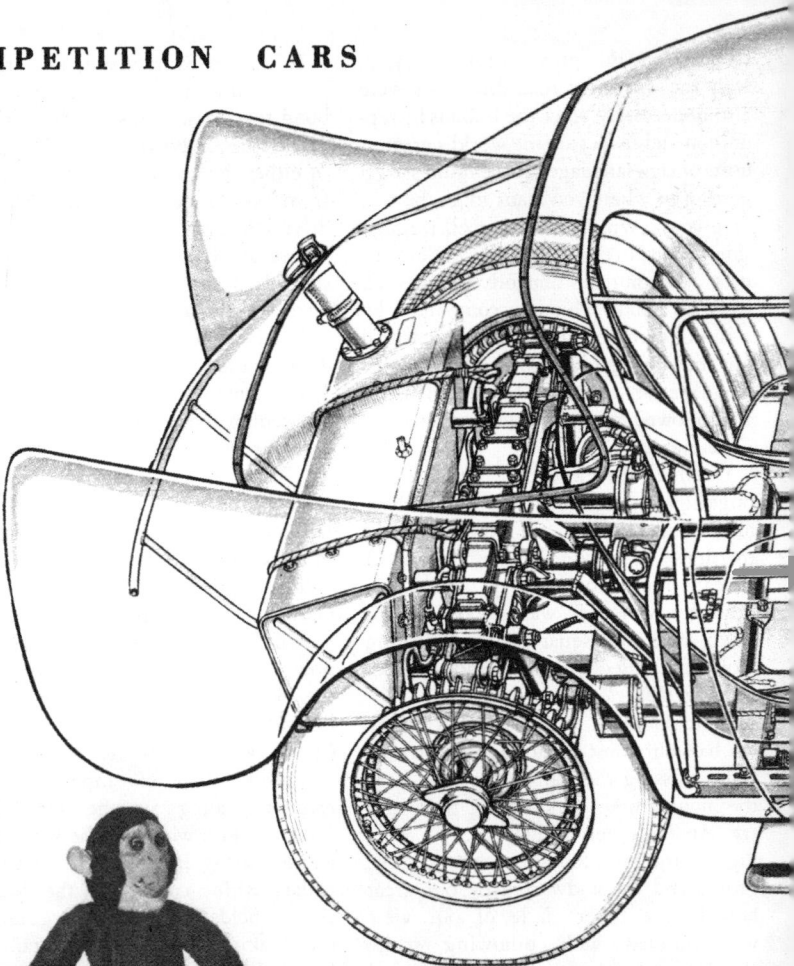

The rear extremities of the main side members are bridged with a box section cross member, and this carries the inboard frame extension tubes; these in turn are welded at their forward ends to a tubular cross member, placed some 10in towards the front of the car. The frame extension tubes carry the final drive unit mountings, to which are bolted the rear leaf spring bridge piece.

Coil springs form the front suspension medium, wishbones being cast in RR 51 alloy; the spread of the wishbone arms is particularly wide to give fore-and-aft rigidity against braking stresses. Standard M.G. swivel pins and stub axles are used. Similar cast alloy wishbones are used at the rear, the suspension being by a multi-leaf transverse spring. This is mounted on a cast alloy bridge piece which is bolted across two welded-up arches of 1½in diameter tubing. The outer ends of the spring slide in trunnion blocks bolted to the upper wishbones, and the bearings are made of duralumin. The rear parts of the arches on to which the spring saddle bridge piece is secured also form the rear mountings for the Salisbury hypoid final drive unit; the nose is clamped to the box section cross member. Andre telescopic dampers are used front and rear.

Steering is by Morris Minor rack and pinion, the steering box being mounted below the radiator on the front cross member. The steering column runs in two dural bearings and a Derrington

M FOR

A Gallay oil radiator is mounted in front of the coolant radiator, also of the same manufacture. Rack and pinion steering and RR 51 alloy wishbones can be seen

The sparse use of tubular framework for the glass-fibre body shell, which has six pick-up
points, can be seen in this cutaway drawing

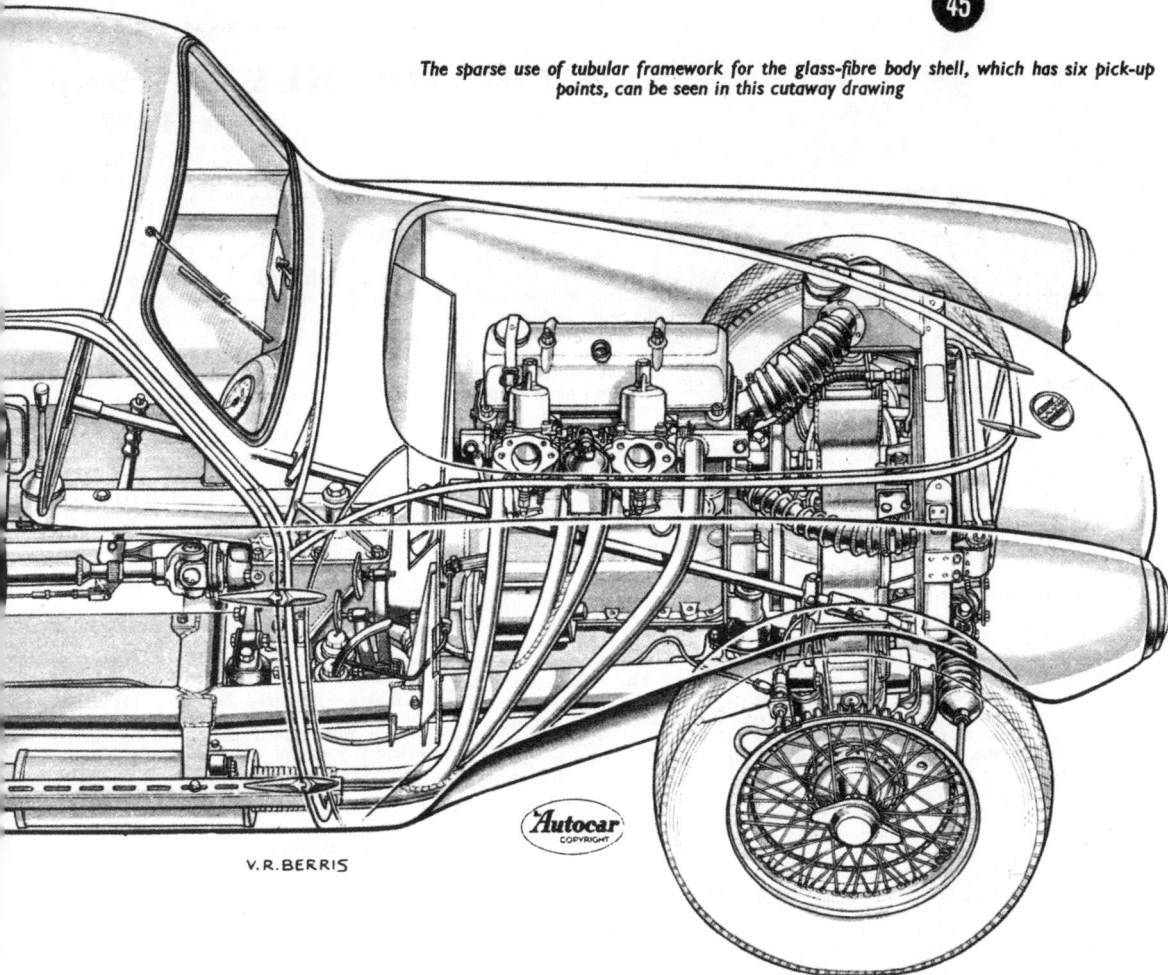

V. R. BERRIS

Autocar COPYRIGHT

MONKEY STABLE

LESTER-M.G. TO RETURN TO THE LISTS IN ENTIRELY NEW FORM

aluminium steering wheel, with wooden rim, is used. Steering arms are by Riley.

The engine is basically the TC M.G. unit but the block is bored out 3mm, making the capacity 1,467 c.c. The compression ratio is raised to 10.5 to 1 and the 1¼in square inlet ports are bored out to 1¾in diameter. Twin 1½in S.U. carburettors are used and the inlet manifold is the Stable's own, fabricated as a welded unit. A separate float chamber, mounted on the body frame, feeds both carburettors. The pistons, connecting rods and camshaft, too, are non-standard, as are the clutch and flywheel, and the cylinder head has received the attention of Harry Weslake. The engine exhausts via a Servais-manufactured system, pipes 1 and 4, and 2 and 3 being paired to two separate inlet pipes into the silencer.

A Purolator oil filter is used in place of the normal Tecalemit and a Gallay oil cooler is mounted in front of a water radiator of the same manufacture. The oil circulates from the sump, via the pump to the filter, through the oil radiator and back into the engine. Wherever possible for parts that are not highly stressed, alloy has been used to keep the weight down. It is clear that, throughout the prepara-

tion of the engine, great care and trouble have been taken—on removal of the valve cover this becomes particularly evident. The rocker shaft is carried on light alloy pedestals; each rocker arm takes four days to make and is beautifully finished.

With these modifications it is claimed that the extremely high output of 100 b.h.p. at approximately 7,300 r.p.m. has

been achieved, and a safe limit of 8,000 r.p.m. has been set.

The gear box is standard TC M.G., being some 10 lb lighter and a little shorter than the later pattern. This gear box will also be used in the Coventry Climax-engined cars, a modified bell housing being adapted to fit. A Hardy Spicer propeller-shaft with two universal joints

Servais-manufactured, the exhaust system permits an extremely smooth gas flow. Twin 1½in S.U. carburettors are used, with a common float chamber mounted on the body frame

M FOR MONKEY STABLE
. . . . continued

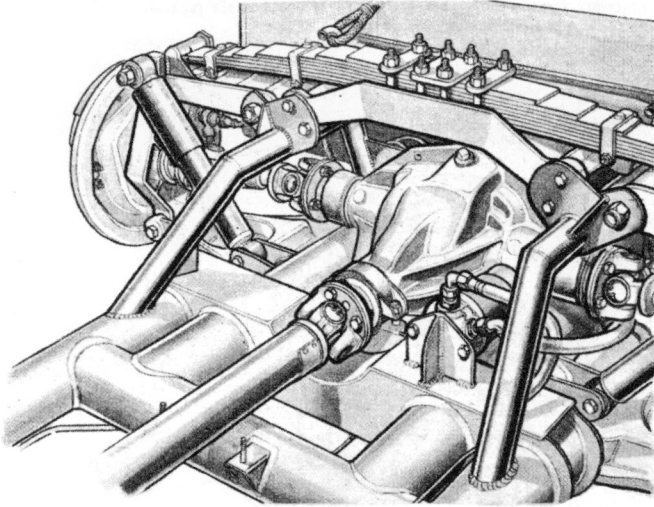

The rear suspension, showing the Andre friction type dampers and tubular members carrying the cast alloy bridge piece for the transverse spring. The large capacity S.U. fuel pump is mounted near the Salisbury final drive housing

completes the transmission. Brakes are Lockheed hydraulic with twin master cylinders for the front and rear wheels, used in conjunction with Al-Fin drums. On subsequent cars Al-Fin Mark II drums, with turbo-fins, will be fitted.

The central, fly-off hand brake is a standard M.G. component and operates by means of cables on the rear wheels only. The rear brake back plates, which are light alloy castings, carry two oil seals, an inner ball race and an outer, taper roller race. The outer end of the short length of shaft running in these bearings is pressed into the hub which, in turn, is bolted to a race carrier attached to the H-bone. By this means, if this shaft should break outside the outer universal joint of the half-shaft, the wheel will still remain in place. The Dunlop centre-lock wire wheels are fitted with 5.50 by 15in Dunlop Racing tyres.

On the first completed chassis, with the 1½-litre engine (the model will be designated the M15), a modern, all-enveloping, glass-fibre coupé has been fitted. This, in life, is very attractive, much more so than the photographs indicate. Glass cloth only (not mat) has been used in the laminations and the thickness is only $\frac{3}{32}$in; the total weight of the untrimmed and unpainted body shell is only just over 70 lb. A tubular steel framework is used to support the shell, which is attached to the chassis frame at six points—two on either side of the cross member extensions, two on the steering box mounting lugs and two on extensions of the rear chassis members. In spite of the light-ness of the body shell it is surprisingly strong. Great trouble has been taken over seating comfort, the seats being specially made with side pads to give lateral support. The 1,100 c.c. cars (designated M11) will have open two-seater coachwork, also in glass fibre.

A 15-gallon petrol tank of Gallay manufacture shares the tail of the body with the spare wheel and is filled through a quick-action filler cap which protrudes through the tail. The tank, which is of light alloy, is secured to the rear extensions of the frame by means of rubber bands and is insulated from the frame by a Sorbo rubber mattress. Fuel is delivered to the float chamber by a high-pressure S.U. electric pump.

The M-type Lester-M.G. is a very workmanlike and well finished, small sports racing car built for a class that, until recently, had all too few contestants. When it appears—as is hoped, for the May 7 Silverstone meeting—it will be coveted by many. Though, obviously, the needs of the team cars come first, it is hoped—hoped, not promised—that there may be subsequent cars available for sale.

SPECIFICATION

Engine.—M.G. four cylinders, 72.0 × 90 mm (1,467 c.c.), o.h.v., push-rod. Compression ratio 10.5 to 1. Two 1½in S.U. carburettors, coil ignition. Approx. 100 b.h.p. at 7,300 r.p.m. Rev. limit 8,000.

Transmission.—M.G. TC gear box. Salisbury hypoid final drive. Overall ratios 4.4, 5.96, 8.63 and 14.92 to 1. Hardy Spicer propeller-shaft.

Suspension.—Independent; front by coil springs and wishbones, rear by transverse leaf spring and wishbones. Andre telescopic friction dampers at front and rear.

Brakes.—Lockheed hydraulic, 11½in diameter by 1⅜in wide, working in Mark II Al-Fin turbo-finned drums. Twin master cylinders for front and rear wheels. Central fly-off hand brake operating by cable on rear wheels only.

Tyres and Wheels.—5.50 by 15in Dunlop Racing on Dunlop centre-lock wire wheels.

Main Dimensions.—Wheelbase 7ft 3in, track (front and rear) 4ft 4in. Overall height, coupé, 4ft 1in (approximately). Dry weight, 9¼ cwt.

Above: The nearly completed glass-fibre body shell of the first car, and in the foreground, the second 1½-litre chassis. Left: The TC M.G. gear box with remote control gear lever. The two reservoirs for the twin Lockheed hydraulic master cylinders can be seen

R & T CLASSIC TEST NO. 3

1949 MG-TC

SPECIFICATIONS

List price	$1895
Wheelbase	94.0
Tread, f and r	45.0
Tire size	4.50 - 19
Curb weight, lbs.	1840
distribution, %	47/53
Test weight	2180
Engine	4 cyl, ohv
Bore & stroke	2.62 x 3.54
Displacement, cu in.	76.3
cu cm.	1250
Compression ratio	7.25
Horsepower	54.4
peaking speed	5200
equivalent mph	80.6
Torque, ft-lbs.	63
peaking speed	3000
equivalent mph	46.5
Gear ratios, overall	
4th	5.12
3rd	6.93
2nd	10.1
1st	17.3

CALCULATED DATA

Lbs/hp (test wt.)	40.0
Cu. ft./ton mile	78.3
Engine revs/mile	3870
Piston travel, ft/mile	2280
Mph @ 2500 fpm	65.6

PERFORMANCE, Mph

Top speed, avg.	75.0
best run	78.9
3rd (5500)	63
2nd (5500)	44
1st (5500)	25
see chart for shift points	
Mileage range	22/28 mpg.

ACCELERATION, Secs.

0-30 mph	5.7
0-40 mph	8.8
0-50 mph	14.0
0-60 mph	21.2
0-70 mph	34.3
Standing start ¼ mile	21.8

TAPLEY DATA, Lbs/ton

4th	170	@	40 mph
3rd	240	@	36 mph
2nd	350	@	32 mph
1st	430	@	23 mph
Total drag at 60 mph, 120 lbs.			

SPEEDO ERROR, Mph

Indicated	Actual
30	30.5
40	39.9
50	49.3
60	58.2
70	68.1
82	78.9

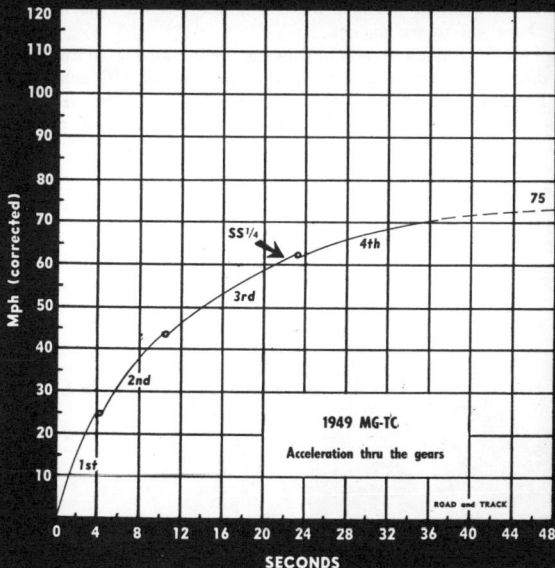

1949 MG-TC
Acceleration thru the gears

ROAD and TRACK

1949 MG-TC

Ed. Note: The following nostalgic report was put together from notes made by the Editor over six years ago and the data was taken from an actual car which belonged to Mrs. Bond.

February, 1950—The advent of the MG-TC on the American market was not accompanied by any great fanfare, but already the completely new idea of using a small sports car for fun driving has captured the imagination of a public normally interested only in transportation. The recent price reduction from $2395 to $1895 should do much to accelerate acceptance of this new way of life.

Certainly the first time we took the wheel of a TC, it felt strange indeed—what with the right-hand drive, left-hand shift, low seating position and ultra-quick steering. The latter point was quite disconcerting at first, and we literally staggered down the street for a few blocks. Very soon, however, one gets the true feel of the car and its charm grows with the miles.

A gearbox that must be used properly, and can be, is something new to us but it soon proved to be sheer pleasure. With a foolproof synchromesh on the 3 upper ratios one anticipates the need of the next moment and employs the British "downshifting" technique. The MG is not designed to be a high-gear performer, but proper use of the gears gives remarkably brisk performance, the engine buzzing away merrily all the while. For a four-cylinder unit the powerplant is remarkably smooth; it will run up to 5500 rpm at least without complaint or worry, and we once touched 6200 rpm in 2nd gear.

The results of the timed high speed runs may cause some "flap" among MG owners, but our data was taken with an ample run-up and agrees with similar reports published in England. We found no important difference in speed between top-up and top-down, but overseas reports say that folding the windshield forward adds 2 or 3 mph. Having no goggles or vizor, we did not try. Suffice it to say that the axle ratio is near-perfect for best possible top speed since the engine peaks at 5200 rpm which is very close to the best run of 78.9 mph.

The quick steering (1.7 turns, lock to lock) has been mentioned, although the turning circle is rather poor at 37 ft. The suspension is very stiff at the rear and the MG corners like the proverbial train—"as if on rails." If the proper gear is selected (high is not powerful enough) a corner can be taken with a beautiful drift with very little "sawing" of the steering wheel. There is a slight amount of understeer which is not objectionable and can be reduced by using relatively low tire pressures in front and rather high pressures at the rear. Unfortunately this treatment is at cross-purposes with the ride characteristics, for the front end seems too soft for the rear (or vice-versa) and the radiator bobs up and down rather too fast over rough roads. High speed stability is not too good and requires alert attention to the steering on narrow roads. One run, down-hill, at 90 mph indicated, proved this—keeping a safe straight course at this speed being almost impossible.

With all its faults the TC will prove to be a dearly loved possession. It is extremely durable and many of the cars owned by our friends have taken really terrible treatment and abuse without complaint or trouble. Styled in the true classic sports car tradition, one hopes that the MG Car Co. will continue this model without resorting to the new fashion of all-enveloping bodywork. The TC has a character all its own and already has taught us that driving can be fun once more. ●

MG's Greatest

---the Classic TC

FOR SOME unaccountable reason there has never been a road test of an MG-TC published in *Road & Track*. Perhaps this omission was due to the fact that at the time we began putting out regular monthly issues, the TC was being superseded by the TD; perhaps also we assumed that the TC was so familiar to our early readers that a formal introduction would have been presumptuous. But these are flimsy excuses, and the hard fact remains that only now—nearly ten years after our first issue—are we finally paying homage to the one car above all others that kindled the spark of the post-war sports car movement in the U.S.A. And during the years of taking the TC for granted, it has slowly changed from contemporary to classic. There are always those who will argue that true classics must date from before the Second World War, but most enthusiasts will agree that the rank of classic for the TC is unassailable, if for no other reason than that it stems directly from the TB model introduced in early 1939.

The illustrious history of the British MG has been told many times in many places and probably nowhere better than in John W. Thornley's definitive book, "Maintaining the Breed." We will not attempt to recount the saga here because we are concerned mainly with one model. It is worth remembering, however, that the very first MG was built in 1923—a lean body mounted on a Morris Oxford chassis and fitted with a pohv Hotchkiss engine. It topped 80 mph and created quite a sensation, but it was not until 1930 that MG's racing history is generally considered to have begun. In 1929 the car's name-initials, which have puzzled so many through the years, were explained as being in honor of Sir William R. Morris and standing for his original company, "Morris Garages," and they have stood ever since suffixed by a bewildering alphabet of model designations to confound the layman and delight the expert. The first MG Midget appeared in 1929, and then began the long series of brilliant racing achievements and record runs which actually have yet to see an end.

Nowadays there is an almost unanimous feeling among sports car people that the TC represents the high-water mark of MG achievement. There are MGs that are faster, roomier, far better riding, quieter—but there has never been a model that stood for so much

Restored in pale blue paint with tan leather, this TC is a steady Concours favorite in Southern California.

photography: Poole

A winning show car must be spotless inside as well as out.

Optional equipment includes tire cover, arm rest, and "trafficators."

TCs came with right-hand drive, but having tach and speedo widely spaced was more symmetrical than practical.

Left side of engine shows a very large 12 volt generator for a very small car.

that is "thoroughbred" about a sports car. It is hardly exaggerating to say that when in the late 'forties the TC made its appearance on the highways of this country, it stood out from other vehicles like Swaps or Nashua in a field of plow-horses. For a comfort-loving public it was wretchedly impractical; your spine was jolted, your knees bumped, you were hot in the sun and wet in the rain, you had no luggage space and only 54 bhp—but for the first time in many a year you were *driving* a car. A person felt that it was *part* of him, as quick and responsive to his commands as a well trained mare, and for many a U.S. driver this was something new and wonderful. Inevitably concessions to comfort (TD), performance (TF-1500) and modern styling (A) had to come, but somehow with each succeeding model the prestige of the TC just continues to grow. As those running acquire more and more miles, finding a really "clean" one becomes more and more difficult. When, therefore, we encountered the TC shown on these pages at a Bakersfield concours d'élégance (where it won its class), we examined it thoroughly, found it just about as "clean" as they come, and decided to present it in Salon.

Mr. Paul Vusovich is a sales representative for Cadillac's factory retail division at 5151 Wilshire Blvd. in Los Angeles, and as such, not the most likely person to be an MG owner. When, however, he purchased this car two years ago for $620, it was not in the best of shape. A minor accident involving a crumpled fender decided him on the restoration undertaking, and, although not as elaborate a job as shown on the next page, his work was sufficient to win a roomful of trophies including best-in-show at the Palm Springs concours last February. Mr. Vusovich's aim was not to have the gaudiest TC possible, but rather an authentic restoration of the car to its original state. The body was not removed, but its baby-blue paint was scrupulously matched and the tan leather refurbished. More than $2000 dollars have been spent bringing the car up to its present condition, but with concours-ready TCs as highly prized as they are today (even a reasonably preserved one can bring from $1200-1500), it is not inconceivable that Mr. Vusovich could retrieve his investment, should he ever decide to sell. At present, however, he is content to bring home an occasional cup and give his wife and five children turns riding around on the weekend.

On the next page you will find a picture story of some of the work involved in restoring a TC if you want to go "all-out." The car has been a frequent competitor to (and once victor over) our Salon TC, but owner V. E. Ellsworth's emphasis is more on competition trim. In either case, there is hardly a sports car of any make or country more worthy of the attention a proud owner can lavish upon it than the MG-TC, and if we seem overly affectionate it might be because we sometimes wonder—without the TC bug to bite people a few years ago, who would have started reading "America's Sports Car Magazine"?

Restored MG-TC of V.E. Ellsworth, Mercury International Pictures, Hollywood . . . *"much rubbing, many man hours . . . . original budget a mere memory . . . . body now has 30 coats of lacquer . . . . tan paint under fenders, runningboards, hood, matches upholstery."*

photography: Ray M. Johnson

Restoring an MG-TC

". . . where did all those parts come from?"

" . . . . removing old paint proved a huge job."

" . . . . begins to look like a car again."

" . . . . body assembled and ready for upholstering."

" . . . . baked crackle enamel dash made from sheet steel."

" . . . . right side of engine . . . . more hand-rubbed lacquer."

" . . . . engine work by Moss Mtrs. . . . . chrome everywhere."

S.C.W. Classics Salon

story: MIKE KABLE

M.G. MIDGET SERIES

T.C. 1949

M.G. MIDGET SERIES T.C. 1949:

BUILD IT YOURSELF — IN NINE MONTHS FLAT

OF the long line of "magic midgets" which emerged from the M.G. Car Company's humble factory at Abingdon-on-Thames, none has enjoyed the extraordinary popularity of the TC — fast becoming one of the really legendary sports cars of all time.

Perhaps the TC's greatest appeal came from its utter simplicity and from that rare vintage quality so seldom found in a relatively modern car.

For one thing, the TC was several years *behind* its time. It was always M.G. practice to supersede existing models after each three years, and when war broke out the TB—the TC's blood brother—had been in production for more than a year.

After the war the M.G. Car Company, forced to run on a shoestring budget, decided to capitalise on what it already had. So the TB came back. With minor alterations, it was tagged with the type-name TC for the sake of progress.

Immediate response was not encouraging. Sales were slow enough to cause the company some concern about its rather hasty decision. But only a few months later the TC was riding the crest of a boom wave which ended only when the model went out of production. Orders poured in from all over the world. The factory literally burst at the seams to keep up with a demand from thousands on thousands of sports car enthusiasts both at home and abroad. The TC was in and would stay that way, it seemed, forever.

Where else? A car that throws you back a mirror reflection from the undersides of its mudguards . . .

Not even Cecil Kimber in his wildest dreams could have visualised the fantastic success of the little renegade from the late thirties.

Of course the day of reckoning had to come. The three-year period was drawing to a close and another "midget" was due to take the TC's place.

Ardent worshippers at the TC shrine greeted the new model — the TD — with mixed feelings, just as those dyed-in-the-wool enthusiasts did when progress reared its head in the shape of mass production and overhead valves in the late 1930's.

Now we've moved to 1959. The TD and TF have come and gone and with them have died the traditional lines of the world's best-known and best-loved sports car. The MGA was a new and exciting experiment, just as the desirable Twin-Cam is now.

There are those who still hang on grimly to their outmoded TCs, and when you discover one as beautifully restored as that owned by young Sydney mechanic Phil Small it's easy to see the reason why. Even a perfectionist couldn't fault this car, because Small himself is a perfectionist.

It wasn't like that a year ago. Small's car lay in a crumpled heap against a set of park gates after a collision with another car. Half the bodywork was torn away and the chassis was badly twisted. An insurance assessor took one look and marked write-off across his pad.

Small had owned his car less than 12 months. It had covered

S.C.W. Classics Salon

M.G. MIDGET SERIES T.C. 1949:

BUILD IT YOURSELF
— IN NINE MONTHS FLAT

less than 50,000 miles and was in good mechanical shape. So he bought it back from the insurance company for £140 and set a target date of 12 months in which to completely rebuild the car.

First, the chassis was despatched to Jack Pryer for resurrection. It was straightened, re-riveted, tempered, crack tested and gussetted along the front arches. Then it was fitted with special wide, tapered-leaf rear springs. Shock absorbers were converted from double to single action, new front springs were mounted and the twisted front axle was de-twisted and bolted into place.

A bargain £12 bought Small a complete scuttle, doors and rear bodywork. Using the old framework as a pattern, friend Phil Boot, of St. Ives, re-timbered the whole car with imported coachwood. This took just on three months while every curve and every join of the original woodwork was faithfully copied.

Small, out of necessity more than anything else, taught himself panel beating and duco spraying. He fashioned some new scuttle panels from heavier-gauge, rustproof body steel and repaired any slight blemish likely to show out in the finished job. The metal was rubbed right back and when the

body went on the chassis it was finished with 14 coats of paint.

Other bodywork replaced included the front mudguards, running boards and apron.

The mechanical shake-up was extensive. First of all the block went to Eric Pengilly at Cammeray to be bored and sleeved to 1,500 c.c. with Laystall liners. When it came back it was fitted with a new oil pump, timing gears, timing chain four-ring pistons, conrods, a three-quarter race camshaft and an ultra-light flywheel from which 10 lb. had been chopped, bringing the weight down from 22 lb. to 12 lb. The crankshaft was reground and fitted with wide Van-

CONTINUED ON PAGE 67

SALON

MG TC 1949

TRYING TO EXPLAIN rationally one's feeling for the MG-TC is somewhat like trying to expound on the difference between a Picasso and a Millet—it's difficult, no matter how hard you try—though there are many obvious and important divergencies in approach, thought and feeling.

Casting aside sentimentality, the only reason the TC appeals is because of its appearance. What else counts—in a car? Any clod, especially a clod, can appreciate a Millet, but it takes something more to feel the fine points of the TC. Or does it? The TC had something that defies description—class, style, appearance—whatever it was, we suggest (humbly) that the TC was a thoroughbred in appearance, if not in price. The epitome, if you please, of an era not so far past as to be forgotten by our elders, and recent enough to be respected by *nouveau* enthusiasts.

PHOTOGRAPHS BY MARVIN LYONS

Appearance, then, is the prime consideration. The TC was long, lean and lithe. It looked fast and racy, in the idiom of the past, regardless of whether it really was fast or not (it wasn't!). But if you gave a good modern industrial designer the job of cleaning up some of the early MGs the net result would be—the TC. Either by accident or intent, the net result was (and is) an almost impeccable automobile design.

Consider, for example, that long hood, with the windshield almost in the middle of the chassis. "Nonfunctional," cry the idealists, but actually if you must seat a man low and just in front of the rear axle—where can his legs go but under the cowl? Naturally, the engine belongs behind the front axle, and the radiator dead-center above. Never mind about polar moment of inertia or K^2/ab ratio—good cornering takes stiff springs, preferably wrapped in strong tape to make them even stiffer and, theoretically, better.

Or consider the matter of wheels. A sports car is designed to be raced, and racing demands wire wheels for strength and lightness, plus the very necessary ability to be quick-changed when every second counts. What is more honest than a BIG wheel, say 19 in., to give the personification of a hairy-legged monster designed to traverse any type of road or track, regardless of its condition—and at any speed?

And the engine. If it won't turn 5500, or even 6000 rpm on occasion, it just isn't even worth thinking about. No matter about the cubic inches; it's the bhp per inch that counts. The big Yanks have their bags of lb-ft, but the tiny engine with the right choice of gears—that's the way to drive a car—and let the driver enjoy it.

Maybe all this doesn't adequately describe the MG-TC (we

ILLUSTRATION BY WM. A. MOTTA

know it doesn't). But the fact remains that no other car, regardless of price, will ever quite have the spot in the hearts of enthusiasts as this big-little car, the MG Midget. And when we see one restored, especially one that is authentic (and not *over*-restored), our hearts tend to function faster than our reason—justifiably, we like to think.

There are some things one just can't explain, and the TC is No. 1.

The TC pictured here is the possession of auto upholstery expert Bill Colgan, of Burbank, Calif. Bill did the complete restoration of his car (including new top and upholstery) as well as the upholstery of Tommy Wolfe's Aston Martin featured in last month's *Road & Track*.

MG
TC 1949

THERE was a time when I believed, in all innocence, that a machine could not have a personality or character. An intimate acquaintance with an ancient red MG TC with 100,000 miles on the clock changed my mind very quickly.

Labor Day weekend some three years ago had been chosen as my grand debut on to the motor racing scene, and for more than three weeks before every waking moment was spent on preparation for the great event.

As neither my mechanic, Alan, nor I had any great knowledge of the vehicle, it became a matter of thinking, hammering, and plastering knuckles. It was not long, however, before we discovered that not all the sticking plaster in the world would cover the gaps in our knowledge.

Somehow we managed to attend to all the more obvious jobs and in the misty dawn of Saturday morning we loaded two tool boxes, a fire extinguisher, two sleeping bags, spare clothes, cooking utensils, a large box of canned food and a small metho stove into the back of the TC.

Then we faced the problems of fitting ourselves in.

Fortunately Alan is rather small, and after a great deal of pushing and grunting all was ready. The TC, however, had other ideas, which she showed to great effect by flooding a carburettor.

After much waving of spanner, and a few choice magic words selected from our automotive vocabulary, the recalcitrant beast burst into life with a reverberating roar. As we sped down the road to a chorus of oaths and banging sidewindows I had a distinct impression of imminent adventure.

Climbing into the mountains, with frosty sunshine sparkling through the mist, and an occasional Jaguar or TR burbling past, all seemed right with the world.

That is it did, until a side-basher Minor came scurrying past with an indecently loud blast from its muffler. Now, being passed by Jags and TRs is all right. Even the odd TF is acceptable. But a side valve Minor? I mean to say, it just isn't sporting. We set off in angry pursuit, but in spite of all our efforts the Minor steadily drew further and further away.

This was too much for the poor overloaded old girl, and as we approached Lithgow the steering became extremely sloppy.

This time she had signified her hurt pride by wearing out the drag link end, causing us to use all our spare steering parts, and our incomplete automotive vocabulary, in order to effect the repair.

It was almost mid-morning before we left Lithgow and as we intended to spectate at Bathurst before going on to Orange, the TC was pressed hard all the way. The closer we approached Bathurst the darker became the cloud of gloom that had settled over the car. Until, as we drove up to McPhillamy Park, the steering exhibited that familiar sloppy feeling.

If there is one thing you cannot buy in Bathurst it is a ball joint for an MG TC.

After visiting most every garage in town, we finally talked a replacement out of the owner of the only square-rigger TC competing at Bathurst that weekend. By the time we had it fitted practice had finished, and to the queries of "did you see Jack Brabham go?" or, "what did you think of Stillwell?" we just had to shrug and look slightly stupid.

Come meal time we found we had left the spirit for our stove at home, and all the firewood to be had had already been taken.

I swear I heard the beast chuckle as we sat there, grubby and greasy, eating cold curried sausages indelicately from the can.

Other motorists, too, had troubles. One of these a chap in a Holden, complete vit vite valls, beaver tail and cushions in the back window, was

having trouble with an out of tune horn. As we all know this is a most serious defect, and if left for any length of time causes a definite deterioration of prestige, and therefore must be rectified at the first possible moment.

It must have been quite serious and nerve-wracking, as it was not until 11.30 that a triumphant blast heralded his success. Peace settled over the Mount for at least 10 minutes before a low rumble told of the approach of a TR3. After the TR had parked, practically on top of us, we received a very interesting lesson in the art of carburettor tuning, as it would appear that the top of Mount Panorama at midnight is perfect for adjusting SUs.

For the next two hours we were treated to full throttle bursts interspersed with calls of: "Pass the quarter five sixteen. Yes, of course I mean Whitworth.", "Blast." and "Damn." And such other hard language.

For breakfast, in the cold, hard light of dawn, we opened a pack of biscuits and a can of baked beans. In desperation we decided to see if we could borrow or steal some hot water to make coffee.

We were feeling rather pleased with ourselves as we returned with a steaming hot billy, but not as pleased as the blue cattle dog that was just chewing the last of our biscuits. Alan made a dive to rescue our beans, and the TC considerately pushed a dumb iron between his legs, sending him sprawling into the dog.

Arms and legs flew in all directions, as Alan and the dog staged an impromptu Can Can over the bean can.

Seeing an opening, the dog dived between Alan's legs and dived straight for me. As I shot sideways the TC struck again, and grabbing the handle of the billy with a knock off hub nut deposited the boiling contents in my boot.

We had canned peaches for breakfast.

The rest of the day was spent as far as possible from the TC.

Brabham drove beautifully that afternoon, and it inspired and urged us to go motor racing ourselves; an urge which we gratified by setting course for Orange at the finish of the program.

The shine left the headlamps as soon as we turned the octagon towards Orange, but in our exuberant state of mind we hardly even noticed. The gloom grew deeper as we approached Orange, finally showing itself in a locked steering wheel. As we were negotiating one of those difficult corners at the time, things became decidedly dicey. Out with the tools again.

The lower bearing had collapsed, and so had our faith and once again we opened our entire vocabulary in an effort to free it. Eventually Alan had a brilliant idea, and threatened to burn the thing where it stood.

Night had fallen by the time we reached Orange where we were told by a local resident that all the brass monkeys had gone north for the winter. We took this as an ill omen, and it was with a devil-may-care attitude that we opened our thinning wallets and went looking for a motel for the night.

It was not long before we found where the monkeys had gone, as there were almost as many "No Vacancy" signs as there were people looking for vacancies.

Finally, as we presented our grimed and bedraggled bodies to a certain motel manager for the third time, a look of pity cracked his granite features, and he offered us the laundry. Standard rates, of course.

We were soon under the shower removing a gelatinous coating of dirt from our skins. We then wrapped ourselves around a couple of steaks.

This made a most remarkable difference; in fact we felt almost human as we went to the scrutineer. If we had known just how determined the TC was, we would have gone home then and there. By train.

The scrutineer was quite decent about the whole thing, and even recommended a small garage where he said we could work on the car in the morning. After all, it would only take an hour or so, and I could then start at the rear of the grid.

It took four hours.

First the bolts jammed. Then we had trouble locating a new bearing. After that we had trouble fitting enough shims.

The local radio station was broadcasting the meeting, and we heard our race run as we cut our umpteenth shim with a pair of blunt nail scissors.

Strangely enough, the bolts did not jam as we put the steering box back in the car. We reconnected everything and turned for home, convinced that racing was for the birds, but determined to try again some day.

On the way home the TC found an extra 1000 rpm. There was not a Minor on the road that could even look at us.   #

THE MALEVOLENT MIDGET

BY ROBERT W. HITCHINS

Right: A TC bouncing into Lanes Corner at Phillip Island. Boxing of the front chassis ends improved the braking, handling and ride.

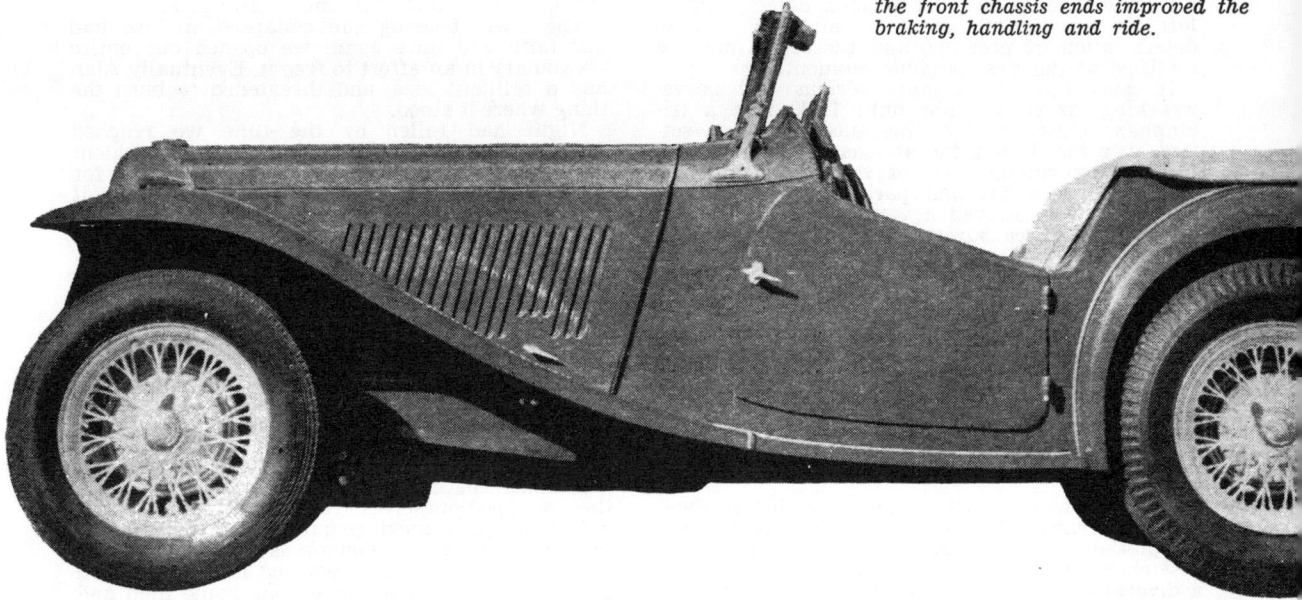

In which a suffering and long-afflicted owner, looks at the problems of making the MGTC into a motor vehicle.

YOU CAN

THE MGTC, born 1946 and still going strong, is one of the most respected little cars in the world. In Australia they started many people on the way to being eternal sports car nuts. Ask your next door neighbor — the one with the E-type — and he'll tell you that his first car was a *TC*.

In their day, they were a remarkable little car. But it's important to bear in mind that a TC was really the *only* sports car available in the forties. With its pre-war design, the TC was one of the last cars built with a front beam axle and wooden body. Nowadays 98 percent of TCs are bumpy, draughty, cramped, slow and rattly. They are covered in chassis cracks and their doors fly open over bumps and they're murder to drive at over 60. To drive from Sydney to Melbourne in a TC is murder too. At 50 mph, a TC is revving away at 3500.

To keep ahead of a Holden or Falcon your TC must have a 9.3 head with bigger valves, extractor and one of those BRM superchargers. And then you'll discover that the sedans will outbrake the TC. On bumpy corners, sedans seem to outcorner TCs. The only things modern cars lack are the rattles and the bumps. A big reason for this is that most of the TCs are on their last legs. As a TC owner who spent a long time rebuilding a TC and then racing it, only to find that all kinds of cars were beating me across the line, I have a number of hints to offer.

There are things to correct on a TC that improve the car without altering its appearance. Sometimes the mods take only a few minutes.

HOW TO STOP A TC BEING BUMPY: These modifications not only stop a TC being bumpy; the cornering speed increases, and the braking improves tremendously. The work was done originally for racing and it just happened that it became a better road car afterwards. The quickest way to improve ride is to fit smaller wheels, but most TCs appear to have 16 in. wheels now; 15

in. wheels are even better. Yet I found that *13 in.* wheels (used for sprints) improved the handling, braking and ride out of all proportion. Thirteens look too small with the standard mudguards. If possible, fit fifteens. When you have sixteens use the small tyres—especially on the front. On the back, avoid the 6.00 size. They make the car feel heavy and they don't leave much room under the guards.

Most TCs have worn shock absorber pistons and rubbers. Telescopic shock absorbers — especially on the front — improve the ride beautifully. And they don't leak oil or requre attention. They're more efficient and you save a lot of weight by throwing the old ones away. Another way to keep the wheels on the ground (to improve the ride and braking) is to remove two leaves from the back springs and one from the front. The springs are softer after this and the car is lower. This mod takes about an hour and you'll definitely notice the difference.

These three improvements have given a softer ride. By stiffening up the chassis the car will be tighter, the suspension will work more and the ride will improve. So, remove the dumb-iron apron and complete the boxing of the dumb irons. Weld a sheet of thin steel over, then cut it off with the torch and clean off the edges. You'll have to reverse the bolts of the shock absorber mounts but that's no trouble.

The standard chassis boxing stops at the cross-member near the radiator. By completing the boxing, filling in and welding around the cross-member, the chassis won't flex so much. This principle also applies to the last cross-member at the back of the car. The front cross-member is most inadequate. After you've boxed in the dumb-irons, cut a piece of square steel tube to fit between the extreme ends of the dumb-irons. Make it as big as possible — something like 1¾ in. square. Weld it in almost touching the original, round cross-member. Then remove the

FIX THE MGTC

old cross-member by undoing the pins that position the spring. Finish the welding on the square tube and tighten the pins with new nylon nuts. The dumb-iron apron will still fit and the car will feel 100 percent better.

HOW TO MAKE A TC BRAKE BETTER: The smaller wheels, the efficient shock absorbers, the stiffened chassis all help to make a TC stop quicker. This is because the wheels are being kept on the ground. Naturally, by using competition linings the braking will improve considerably. The pedal pressure will be higher but it's worth it (believe me). If you're really serious, the pedal can be bent to allow heel-and-toe movement. It's a natty thing to have, but there seems little use for it around the streets.

Check the rubbers and pistons. I found that my car was wandering violently and the hand-brake cable was jamming, which meant that one wheel was braking continually. Also, it pays to have the eyes and rubbers of the main springs perfect. Otherwise the car seems to slew under braking.

HOW TO STOP THE TC RATTLES: If you drive a TC without mudguards and bonnet, there will be fewer rattles. So, it's important to tighten the mudguards. Very few TCs have the round rubber spacers holding the back guards from the petrol tanks. (Petrol tank: make sure the tank is mounted on rubber.) The bonnet should always be tight, and the rubber strips at each end must remain intact. Otherwise the bonnet will rattle, vibrate and wear away the body.

The radiator grille will rattle if it's not looked at. And what's more, it will usually crack around the hole where the starting handle goes. This can be brazed and reinforced from behind. Often this can be done without having the shell re-chromed afterwards. While the welder's on the job, remove the louvres and have them spot welded. Normally they're riveted and they rattle badly.

After tightening every nut and bolt, the TC will still rattle. It's the body. By shifting the heavy battery from that high position in the scuttle, and putting it behind the seat next to the differential, half your troubles will be over. It may be necessary to have a smaller Fiat battery, but you'll find the results amazing. Weight distribution is improved and you can stow things (like bottles) in the old battery compartment. With the battery away from the scuttle, the body will hardly vibrate.

The windscreen, being high and heavy, moves even when it's slightly loose. It can be tightened properly by fitting longer bolts with nylon nuts. One of the bad features of TCs are their weak body mounts. The front mounts always crack; I have seen only one or two mounts without cracks. They can only be continually welded, and kept tight.

The other thing that usually breaks are the alloy gearbox mounts. These can be welded and I believe the secret is to tighten the surrounding bolts every thousand miles. The gear lever sometimes vibrates. I think the main causes are the weakened selector springs.

Tools loose in the tool box seem to make a hell of a racket. The tool box, being at car level, transmits all the 'clinks and clunks'.

Spring shackles and worn rubbers mean that when a bump is struck there's a loud metal-to-metal contact. So they should be checked.

The steering column rattles and bangs because it sits in a felt bush that wears and moves every time the wheel is turned. Half an inch of garden hose squeezed past the spline makes a bearing that doesn't move and lasts for years. That stops a bit of noise. Oil pipes and cables that run through the bulkhead will be noisy if they are not surrounded by rubber.

Also, I must say that after changing from the old 19-inch wheels, and fitting decent shockers,

A TC on the old Port Wakefield circuit: It is getting wheelspin in second, due mainly to rear 17 in. tyres.

the rattles diminished greatly.

HOW TO STOP THE TC WANDERING: A lot of people say that TCs wander because of their clammy mudguards. And while TCs with cycle types seem nicer to drive, I can't help thinking that a brand-new TC would not have wandered at speed. Besides, the TCs I have driven with cycle types, all had 16-in. or smaller wheels. It's very hard to keep 19-in. wheels in perfect balance. I find that every 12 months or so, there are a half a dozen spokes loose or broken. For a TC not to wander, the wheels, suspension and chassis must be in top condition.

With the beam axle up front, it is important for the spring shackles not to be worn. The hardened steel pin that holds and lubricates the front eye wears fairly quickly. I lubricate it every 500 miles. Wear can be detected by levering between the spring and the dumb-iron. If there's any movement, the chances are the car will wander. This is another reason I think it's important to stiffen the front of the chassis. And I am sure that loose wheel bearings cause the TC to wander. Closely associated with the wandering is the problem every TC owner has: steering play.

HOW TO IMPROVE THE TC STEERING: I once drove a TC in excellent condition — it was fast, had a Laystall head — and the body was perfect. The owner had got over the famous problem of loose steering, by installing rack and pinion. The steering was light and there was no play in the wheel, but the whole column and wheel vibrated tremendously. After that, I drew the conclusion that a beam axle and rack and pinion just couldn't live together.

To have nice TC steering, the first thing is to have nice splines and tight wheel bearings. Then it will pay you to spend the afternoon tightening all the ball joints. If they're too tight the steering will be unbearably heavy. If the balls and pads are worn, replace them. To adjust these joints it's necessary to remove a split pin and turn the adjusting screw and re-align the slot with the holes. Usually, when the adjustment is correct, the slot does not line up. It pays to drill *more* holes so that you can have good ball-joint adjustments without movement.

The steering box has its play adjusted by shims. Now this is crazy, and I could never

work them. Why not drill a hole in the top plate and weld a nut over it? Wind a bolt into the nut (with a lock nut) and rest a ball bearing on the steering 'peg.' So instead of shims you have a simple one-spanner adjustment. The only pressure is on a small section of the ball bearing. This makes the steering accurate and light.

There is often movement on the splined shaft that leaves the box. The drop arm bolts on to the spline and most TCs have the arm loose. I have found that the way to keep it tight is to use a high tensile bolt with a nylon nut and a star washer. Everything else becomes loose after 100 miles.

Now, assuming the box is in good condition, the steering should be almost perfect. To keep the box in good trim and the steering even lighter, clean out the box and remove all grease. Drill a hole in the top side of the column above the level of the box. Weld a nut over the hole and weld the column to the box so that no oil can leak. Pour oil down the hole where the nut is, and when it fills, you can be sure that the oil is above the level of the box. Use SAE 50 oil and the steering will be a bit lighter than when lubricated with grease.

By carrying out all these modifications, the TC will be a nicer car to drive. Without even touching the engine, the car should perform better. It should be much, much safer too. But here are 10 points to keep you thinking:

1. Have you noticed that all TCs for sale are 1949 models? To check this, the month and year of manufacture are stamped on the fuse box and the generator.

2. Unless you live in Darwin, there seems no point in using a fan. My car never boiled. I might add that I also removed the thermostat and the radiator was in good condition.

3. Check underneath the petrol tank. Almost every TC (sorry boys) has a leak. Also, when your petrol light comes on continually, like mine did, you'll know your tank float is punctured.

4. When tightening the oil pump, don't be too strong. Many TCs — including mine — have broken studs.

5. When roaring off in clutch starts, use first gear for an instant. Only snatch first, and your times will improve. CONTINUED ON PAGE 38

CONTINUED FROM PAGE 54

dervell bearings. Every moving part was balanced.

The head was re-worked, ported and polished, and fitted with bigger valves and 1½-in. S.U.s. Some gearbox components were replaced, while the standard 5.25 : 1 differential went out in favour of the 4.8:1 diff. as used in the TA series M.G. Tapered adjustable front hub ball-races were used in place of the ordinary, non-adjustable type and the brake-drums were shod with aluminium Monaro fins to assist cooling.

Finally all the electrical gear was overhauled, wheels were respoked and painted and the car got a re-trim in red vynex. The black vynex hood from the old car was refitted. The job had taken Small just nine months. Expenditure was a little over £300, but hundreds of man hours were poured into the car by Small, Boot and Small's brother, Brian.

As a restored M.G., Small's TC would surely have no peer. The insides of the mudguards gleam just as brightly as the outside. On the road the car has a wonderfully new, taut feeling that the elements of time and wear have removed from most other TCs.

Compression is somewhere over 10 to 1 and Vacuum II racing fuel is used. The engine revs freely and happily and Small hopes that with the addition of lightened rocker gear it will spin easily to 7,000 r.p.m. Some competition work is on the programme, though it will probably be confined to acceleration tests and hill-climbs.

Sitting behind that long, rakish bonnet, bold chrome headlights and upswept guards, it's easy to recapture the thrill that the M.G. inspired in me as a child. I used to make-believe, years ago, sitting in an imaginary midget tearing along an imaginary curving road, tugging that stubby lever back and forth through the gears.

Looking back, I realise that it's probably been somewhat like that right through. My own well-worn TC still fascinates me as much as it did five years ago and enables me to understand why so many apparently sane, level-headed men spend so much time in the restoration of what was originally intended only to be a bread-and-butter sports car for the masses.   #

TEXT: WARREN WEITH. PHOTOGRAPHY: GENE BUT[

MG/TC

THE MG TC TAUGHT A WHOLE GENERATION OF DRIVERS WHAT AUTOMOBILES ARE ALL ABOUT. NOW A NEW CROP OF CONVERTS IS SPREADING THE TC MESSAGE.

• There is a secret country to which no airline flies. Its borders are ever safe from the likes of me because they are guarded by Youth. No one over 30 can break through these defenses. Why mention this wonderful place then? Well, it produces a car . . . sort of the national vehicle. Of course, it's not exportable. You can't even buy one for money.

I was eligible to buy one yesterday. Yesterday was 1949. Sitting in the cafeteria of Fordham College, drinking one black coffee after another, the MG-TC made itself known to me via the pages of *LIFE* magazine. While arguments swirled around my head as the best way of getting through sch and on with the job of building the br new postwar world, I was riveted by article about a group of Californians drove spidery little cars. Odd people. T had formed a club based on mutual in est in these funny machines. Club memb went on rallies (what's a rally?), dr their cars in races—in *races!*—and did sorts of nutty things. Proof was a pictur a club member driving his car under one those high-rise lumber carriers. This to show how low a "British-built spor car" was. Sporting car . . . that ha certain ring to it. They had a look to t that was different. Who made them? N Never heard of that company. Sure m the old gray Plymouth look fat. Made look fat too. Wonder where I could one here in New York?

The under-30 underground worked as well then as it does now. It took ab two weeks to locate an MG-TC owne was Bob Deshon, and he lived in N Rochelle. This . . . this *operator* . . . he cornered his parents into buying him

ost-new example of a "British-built
rting car." By the time I got around to
ting him, and proving that I wasn't just
ging around because he owned an
, the car had been run in the Linden
J.) Airport race. As a result, it was sit-
in the family garage stripped of fenders
headlights. It was suffering from a bent
rod, blown head gasket, and a burnt
e. To me it looked like the greatest
ntion since the wheel itself. It was so
... and those skinny tires were so high!
climbed into it like a World War I
ter plane. It was great. Nothing self-
pelled before or since has gripped my
gination like that poor, over-driven,
nderstood TC.

What was it that I was looking at? Well,
as a way of life. A wildly different car
you jazzed around in on week days
raced on weekends. A mobile set of
dentials that admitted you to a very
ct group. A moving spot of color on a
drab postwar landscape. It could have
n powered by compressed air, or
ssed bubble gum for all I cared. I simply
to be invited to this party that was just
ing.

What was it that I saw those many years
? What it really was was a tiny two-
er marked by a truncated tail and long
ing fenders (when they were used), all
g between four seemingly too-high,
-spoked wheels. Power came from a
-cylinder, 1¼-liter engine. In the cold
of 1967, it doesn't seem possible that
orsepower was all the power there was.
all, thin lump of cast iron, the engine
d in the forward section of a long,
ow engine compartment with foot wells,
als and things taking up the rear section.
river drove in a bolt-upright position
his eyeballs just above the top edge of
cowl. The seat was indescribably awful
t a bucket by any stretch of the imagi-
on—yet we sat in it and sneered at
t we used to sit on in American cars.
ride was hard, but what the hell—so
e the drivers. The gearbox, however,
easy. Stirred by a stubby lever, it
ht a whole generation of road-racers

and would-be road-racers how to execute
heel-and-toe downshifts. The engine hooked
to this box was game, but generally had its
back to the wall trying to fight off some
happy nut who thought he was a blood
brother of Tazio Nuvolari. Usually the engine lost the good fight with a twang and
thump of bent, tinny pushrods ... despite
two decades of auto writers saying things
like, "The MG is as defiant and as tough as
a Staffordshire pit dog." They also talked
about "cornering on rails," those writers.
Only Salvador Dali could have constructed
the rails on which the TC cornered. The
steering was brutally stiff and totally un-
responsive, and the wire wheels were always out of tune—more like potato chips
in their contour. And no matter what—not
even with the Tompkins roller steering kit—
no TC ever proceeded down a straight
road in a straight line. But you learned to
live with the wandering from side to side,
and after awhile you didn't fight it anymore.

continued on page **71**

A TRUNCATED
TAIL, TALL, THIN
WHEELS AND
LONG, SWEEPING
FENDERS
ARE THE TC's
ESSENCE. THE
RIDE WAS HARD,
BUT SO
WERE THE DRIVERS.

continued from page 69

At any rate, Bob Deshon's TC had led a dog's life. I hung around acting as tool passer and gopher (you know, "gopher some cotter pins. . ." "gopher some gasket shellac. . .") while it was put back together. Trial runs without fenders or windshield around the icy roads of Orient Point didn't even start to cool my passion for the TC.

Watching as Briggs Cunningham screamed a supercharged TC down the hill at Watkins Glen further fanned the flames. "My God, when Alfred Momo has got everything screwed on right I can do almost 100 mph!" Breathing Castrol R fumes from the edge of a country road in Bridgehampton didn't help either. The plots and plans I went through trying to buy a TC would have put Machiavelli to shame. I even went so far as to talk some nice young girl—who had a job—into co-signing a note for me. Even that failed. The bank was a little too cagey to put up money for a car that only held two. But those wire wheels . . . that slim-rimmed, quivering steering wheel . . . those long shafts of light bouncing down a blackened country lane. Oh dear.

Being thwarted in my plan to become an owner/driver, I did the next best thing. I hung around with gents who were. I sat around and watched while people like the Collier brothers, John Bentley, Paul Hee, and dozens of others tore down and field-stripped all the MG models —from TC through TD, all the way to TF—and then put them back together again. I read underground magazines, printed on chocolate-colored tissue paper in faraway places like California, that went on for pages about what you could and couldn't do with an MG. I sent to England for pukka, hard-cover books on the same subject. I spent almost enough on books and magazines to buy one of the little beasts. Three-cylindering war-weary MGs back home after the fray got to be a hobby with me. Better than no MG-ing at all. I didn't realize it then—as I pushed one old, high car after another to race meetings, club meetings, midnight garage post-mortems —that it was the best time of my life. That's the way it's always been with MGs, though.

For example, MGs were the best time of Cecil Kimber's life. He was hired by Morris Garages as General Manager in '22. He helped build the first MG special with his own hands. He went on to build the MG company into the biggest all-sports-car factory in the world. Despite a leg badly shattered in a motorcycle accident, he drove like mad and was on hand at every important MG win in the Twenties and Thirties. He worked his way up to such a height in the Nuffield empire that he was one of the few people who could successfully argue the doughty old Lord from a "no" to a "yes" decision. Kimber never became a captain of industry. His salary was always modest, but he knew almost all of the early MG owners by name. You couldn't bank that sort of thing, of course. But you could live one very interesting life, because they were a wild crew.

And they still cast their magic spell. Every morning as I clamber into my square-cut sedan, a neighborhood girl mini-skirts it into a battered red Midget on her way to work. I usually wait to see her get under way. Every morning that Midget . . . with flapping top and shiny-haired driver . . . clears its throat once, then settles down into a crackling idle. In the next minute it's thrust into the traffic stream with a prolonged burst in first gear. She always wears a light smile as the Midget goes by with a lunge into second. And it's not just the newer models that talk a secret language, understood only by the young . . .

The 17-year-old TC whose color pictures haunt pages 67-68 spoke very clearly to David Laemmle. Youthful, in college, with a wide range of modern sports machinery to pick from, David Laemmle had to have this TC. A doctor had restored it for his son, and now it owns David. "It's, well—different—you know? When you drive by, people notice you. Girls are interested in it too. It goes—it'll do about 90—but I don't like to abuse it. Beat a Spitfire the other day. I've put it up for the winter . . . too much salt and stuff on the road." It was like salt in an old wound, listening to that very nice young man talk about his TC. I could go over and help him fiddle with it. Put oil in the SU dash pots . . . things like that. Or now that I've got the money, I could buy one. Except they probably don't sell them to people over 30. Think what I'll do is wait a little while longer. I know a nice young kid who's going to need a co-signer for an MG in about 10 years. Won't he be surprised when he finds out I know how to replace a bent pushrod. I know I'll remember, because the 1977 MG engine won't be all that different. And a good thing, too, in a world that's always changing.   **c/D**

R&T CLASSIC ROAD TEST

1949 MG TC

When the little roadster whipped into the turn ahead of your Plymouth Club Coupe, you knew there was something you'd been missing

GORDON CHITTENDEN PHOTOS

IF THERE WAS EVER a classic sports car, that car was the MG TC. Introduced in England exactly 25 years ago, it was a slightly wider and roomier successor to the pre-war TA and TB models, and before that the visually similar but smaller-engined PA and PB Midgets. To the English motoring enthusiast the TC was merely "maintaining the breed" in satisfactory fashion, but to an American looking down from the heights of his Plymouth Club Coupe it was a fresh, new thing, the very embodiment of "mistress" and an attainable one at that. The TC looked like, and was, fun.

The appearance of the TC was probably as much responsible for its popularity as anything it was actually capable of. The proportions of the 2-seater were as nearly perfect as could be imagined (even our resident styling expert can think of no significant way to improve upon them) and for all the practical advantages of more modern MGs, not a one of them has the beauty of the TC. And it didn't matter, in 1947 when the first TCs were imported, that the concept was already 15 years old. The sweeping fenders, the tall wire wheels, the long hood (more than twice the length of the little 4-cyl engine it housed)—all were just right. Even with top up—that most disastrous of esthetic conditions for a roadster—the TC was extremely satisfying.

But did it really run and handle well enough to justify its 1949 reputation? In that it delivered entertainment, the answer has to be yes. But even making allowances for 20 years of technical advancement, the TC was no model of chassis design and we almost think that the Plymouth Club Coupe could have stayed with it in the corners if the Plymouth driver had had the courage to endure the body lean. But the point is that the Plymouth was work to drive in a sporting manner and the MG was fun.

Sometimes fun is putting up with something you don't have to. Like rock-hard suspension. You can't bounce a TC's front end up and down by jumping on the bumper; it's too stiff. With a beam axle, semi-elliptic leaf springs and lever-action shock absorbers, the TC's front end harks back to the early 1930s, with riding comfort to match. The live rear axle is similarly suspended and the result is a stiff, predictable car that tracks well through smooth corners. We didn't get a lateral-g reading for the TC, out of respect for its age, but the narrow tires produce neither a sizable contact patch nor a sticky one. In responsiveness the TC excels, with the very definition of direct steering, at 1.7 turns lock-to-lock. Steering effort is very, very high and in the first few miles of experience it's hard to believe that the MG was fun to maneuver through tight corners and heavy traffic. But when you get used to throwing a little shoulder into it, it *is* fun.

The cockpit layout is partly to blame for the steering effort. Although the column is adjustable both fore-and-aft and up-and down, the large 17-inch steering wheel—mounted on the right, of course—is too close for proper leverage (Nino Farina, master of the Italian arms-out technique, would never have approved). But the very compactness of the interior does permit everything to "fall readily to hand" (those very words appear in an MG ad in the Vol. 1, No. 1 issue of R&T, June 1947—did MG coin this famous phrase or did they pick it up from some unsung English motoring journalist?). One thing that really falls to hand is the excellent, stubby, short-throw gearshift lever, which is perhaps the most pleasurable part of the TC to use.

The word "snick" may have been invented for the happy sound the box makes going into gear. First ratio is not synchronized, but how can we fault that?—*nobody* had a synchro low in the late 1940s. The tachometer, which includes a small clock, is large and positioned right in front of the driver where the 5200-rpm peaks can be easily read. In the central part of the voluptuously curved, leather covered fascia are the oil pressure, water temperature and generator charge instruments, along with starter, choke, horn, lights and minor switches. Way off to the left, definitely a poor relation, is the speedometer. It would be unkind to say this is just as well with the TC's low maximum speed of 73 mph, when it can be the most exhilarating possible 73 with the wind whipping in through the cut-down doors, or even straight into the eyes with the windshield ⇒

MG TC

ROAD TEST
1949 MG TC

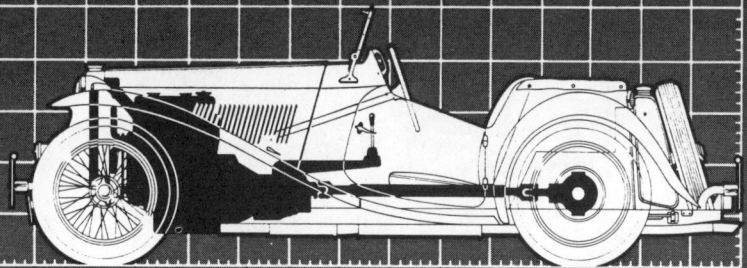

SCALE: 10" DIVISIONS

PRICE

Price $2395, later.........$1895

MANUFACTURER

MG Car Co., Ltd., Abingdon-on-Thames, Berkshire, England (a member of the Nuffield Organization, later absorbed by BMC and now a part of British Leyland).

ENGINE

Type..............4-cyl inline, ohv
Bore x stroke, mm.....66.5 x 90.0
  Equivalent in.......2.62 x 3.54
Displacement, cc/cu in..1250/76.3
Compression ratio.........7.25:1
Bhp @ rpm........54.4 @ 5200
  Equivalent mph.............81
Torque @ rpm, lb-ft...64 @ 2700
  Equivalent mph.............47
Carburetion..........two 1.0 SU
Type fuel required......premium
Emission control...........none

DRIVE TRAIN

Transmission........4-speed with synchromesh on top 3 ratios
Gear ratios: 4th (1.00).....5.12:1
  3rd (1.35)..........6.93:1
  2nd (1.97)..........10.0:1
  1st (3.38)..........17.3:1
Final drive ratio.........5.12:1

CHASSIS & BODY

Layout....front engine/rear drive
Body/frame: separate ladder frame, steel body panels over wood & metal
Brake type: 9.0 x 1.5-in. cast iron drums
  Swept area, sq in.........170
Wheels....wire knockoff, 19 x 2.5
Tires...........Dunlop 4.50-19
Steering type........Bishop cam
  Turns, lock-to-lock.........1.7
  Turning circle, ft.........37.2
Front suspension: beam axle, semi-elliptic leaf springs, lever shocks
Rear suspension: live axle, semi-elliptic leaf springs, lever shocks

ACCOMMODATION

Seating capacity, persons.......2
Seat width.............2 x 17.5
Head room.................39.0
Seat back adjustment, degrees..10
Driver comfort rating (scale of 100):
  Driver 69 in. tall...........60
  Driver 72 in. tall...........55
  Driver 75 in. tall...........55

INSTRUMENTATION

Instruments: 6500-rpm tachometer, 105-mph speedometer, water temperature, oil pressure, generator charge
Warning lights: lights, fuel level

MAINTENANCE

Service intervals, mi:
  Oil change..............3000
  Gear lube change.......6000
  Filter change...........10,000
  Chassis lube.............500
  Minor tuneup...........7500
  Major tuneup..........12,000

GENERAL

Curb weight, lb............1845
Test weight................2190
Weight distribution (with driver), front/rear, %....45/55
Wheelbase, in..............94.0
Track, front/rear......45.0/45.0
Overall length............144.5
  Width..................56.0
  Height.................53.2
Ground clearance...........5.9
Overhang, front/rear....18.5/32.0
Usable luggage space, cu ft....4.5
Fuel tank capacity, U.S. gal...16.2

CALCULATED DATA

Lb/bhp (test weight).........40.5
Mph/1000 rpm (4th gear)....15.5
Engine revs/mi (60 mph)....3850
Engine speed @ 70 mph....4500
Piston travel, ft/mi.........2280
Cu ft/ton mi..............78.5
R&T wear index..........88.2
R&T steering index........0.63
Brake swept area sq in/ton....155

ROAD TEST RESULTS

ACCELERATION

Time to distance, sec:
  0-100 ft.................3.9
  0-250 ft.................7.0
  0-500 ft................11.0
  0-750 ft................14.7
  0-1000 ft...............18.0
  0-1320 ft (¼ mi)..........21.8
Speed at end of ¼ mi, mph....61
Time to speed, sec:
  0-30 mph................5.8
  0-40 mph................8.8
  0-50 mph...............13.8
  0-60 mph...............21.2
  0-70 mph...............41.5
Passing exposure time, sec:
  To pass car going 50 mph..9.5

FUEL CONSUMPTION

Normal driving, mpg..........22
Cruising range, mi...........356

SPEEDS IN GEARS

4th gear (4725 rpm)..........73
  3rd (5200)..............60
  2nd (5200)..............41
  1st (5200)..............23

BRAKES

Panic stop from 80 mph:
  Deceleration rate, % g.not taken
  Stopping distance, ft..not taken
Pedal effort for 50%-g stop, lb..90
Fade test: percent increase in pedal effort to maintain 50%-g deceleration rate in 6 stops from 60 mph..............off scale
Parking. Hold 30% grade?.....yes
Overall brake rating........poor

SPEEDOMETER ERROR

30 mph indicated is actually...29.7
  40 mph..................39.2
  60 mph..................58.6
  70 mph..................68.0
  Odometer, 10.0 mi.........9.8

ACCELERATION & COASTING

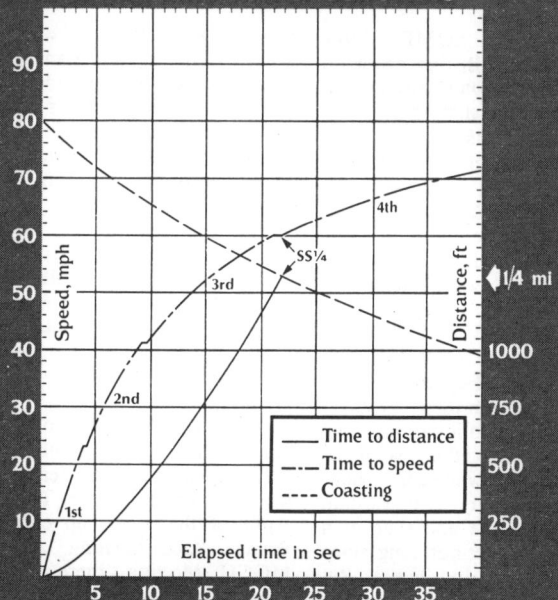

Legend:
— Time to distance
—·— Time to speed
---- Coasting

Speed, mph / Distance, ft / ¼ ml / Elapsed time in sec

CONTINUED ON PAGE 83

THERE'S SOMETHING incredibly British about an old MG. Somehow, upright radiators all in shining chrome, screen raked flat or replaced with aero-screens, drop-away doors exposing elbows, slab-backs with tanks strapped on, goggles and the wind in your hair — all stirring stuff of yesteryear. Yet it wasn't long ago really. And even today, it is possible for pretty incredible performance to be extracted from the legendary T-type — faster now even than in their hey-day.

It is not the policy of this mag to venture into the realms of obscure three-wheeled, chain-driven antiquity, Edwardian taxis, or the life and time of a vehicle manufacturer long since buried in motoring tomes. But anybody who digs cars, performance motoring and can appreciate enthusiasm in preparation must go overboard, as we did, when looking at Paul Langdell's MG TC. A 1947 Bolide that really works. And what's more, we were privileged enough in the bleak mid-winter to canter this 15 cwt yellow carriage of character round Silverstone.

The tale of the TC began when he bought it in '66 in derelict condition, the hulk being very much a tow job. In fact the complete strip-down took three years to complete, working every weekend as well as any spare evening.

At the time he had no plans to Race it, and at the beginning of '69, the car could be termed a standard T-type runner. Then in April of that year, up came a Brands Race Meeting. At the MGCC Meeting, there was a Race for T-types and Paul couldn't resist driving his car from home (to run it in!), Racing it, finishing the Race, and even driving it home again in one piece afterwards. He had caught the bug.

He had a go in a further eleven events during the year. His excuse was merely to 'suck it and see'. Suffice it to say, the car became progressively modified for its extremely active second childhood.

First thing was to reduce weight. The standard all steel bodywork was retained however, still affixed to its ash framework. But off came the original flared front wings, plus running boards, and on went lightweight cycle-type mudguards instead.

A secondhand C75B Shorrocks blower was obtained. The crankshaft was balanced, flywheel lightened, cylinder head polished, though standard valves were retained fitted with 150 lb springs, whilst an extractor exhaust manifold improv... things no end. Head plan... upped the compression ratio... 9.5:1, whilst the capacity w... increased by a 0.100 in overb... to 1350 cc.

The result of all this Lang... tweaking was a transformat... from the 54 bhp of the 1250... motor to approximately 100 ... instead. Thus equipped, P... won a handicap Race at the ... Car Club Silverstone Meeting... '69 being awarded that Clu... Newcomer's Award. All t... with the 16 in wheels still s... with roadgoing Cinturatos.

However, during the win... specialisation reared its u... head, luckily in this case with... meaning a departure from ... car's original character, for a ... of 15 in diameter ex-Cobra ... wheels were obtained, shod w... Racing Dunlops of 5.50 L-...

MG
MAGNIFICENT

RH-E really got down to the grass roots of fresh air motoring when he managed to get his quivering limbs behind the wheel of Paul Langdell's superb MG TC, which turned out to be a real Goldie Oldie.

PHOTOS: SPENCER SMITH

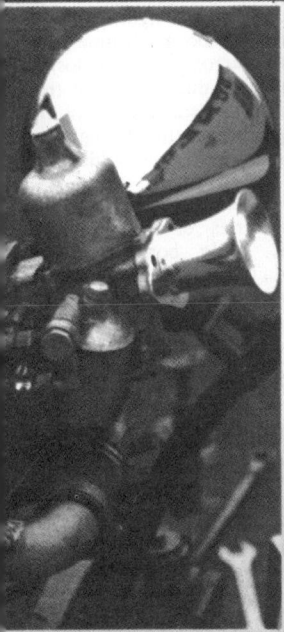

front and 6.00 L-184 at the back. Next, the car was lowered all round, by fitting blocks under the already stiffer springs. During the '70 season, anti-tramp bars were fitted at the front to stop the wind-up of the springs under braking, being home-made using Hillman Minx steering drag links as bases.

With Racing rubberwear, Paul also started to use his brakes properly, thus unearthing more deficiencies in the standard package after such a dramatic power increase had happened. In any case, the new big wheels were shrouding the brakes excessively. Or was it that Paul was driving faster too?

Cooling ducts were fabricated for the front brakes as well as the substitution of the standard linings for harder Ferodo AM4, a servo being shunned as being decidedly cissy. Holes, as aids to cooling, were cut into the back plates.

Transmission always has been standard, apart from a change over to a lower crownwheel and pinion ratio of 4.875:1.

The modifications to the front suspension have proved to be a great success. As original, the T-type owner gets 3 degrees positive camber as standard, with the steered wheels tracked up to be straight ahead. This is all very well for narrow road tyres. But isn't so good with wide Racers.

But after much heating up of the axle tube with a welding torch, Paul and his brother John (swinging on the end of the bar!) changed things according to a home-made jig. This bending has resulted, with half a turn of lock applied, in the outside front wheel now being upright without the steering sensitivity being affected. The castor angle on the king-pins has also been increased from $5\frac{1}{2}$ degrees to 8 degrees at the same time as the camber has been reduced. This castor change is of course necessary as with the camber change on its own, the self-centring of the steering would be affected.

During '70 Paul plus MG TC took part in 21 Races and seven Sprints/Hillclimbs, winning six times. In fact during such a hectic season, only one retirement was experienced and that was over 80 laps in the 750 MC/CCC 6 hr Relay Race when a clutch withdrawal pin broke.

Once unloaded from its trailer, this immaculate conception was cranked into starting,

being warmed up with the aid of an old sock held over the SU's trumpet. Our gloating was interrupted when Paul handed the controls over to us and we were allowed to sample the delights of such a car for ourselves. At this stage it's as well to try to put a value on such a personal car, all very difficult, but in the States in particular the price ticket could be expected to be about £1000.

Out on the circuit, we found we had to exercise great care with our footwork, otherwise the

PHOTOS: SPENCER SMITH

Shorrock blowing device gives a touch of the screaming ab-dabs. Protex-clad Paul Langdell and brother John — upright rad lovers. Scoops, brakes for the cooling of.

result was a series of fits and starts. But it was all there in front of us, the peak of the radiator, the chrome glistening on the headlamp tops, a flat bonnet quivering away like a dumbell in the hands of a novice, whilst the tonneau cover flapped away at our crash hat sides, like someone knocking down doors at the sides of our tester's head.

Gears were a cinch to engage. Not what we expected at all, with a 24-year-old gearbox!

We were most assured to notice that a piece of heather had been taped on to the propshaft tunnel as some sort of quaint charm for a safe passage. But less comforting were the absence of fire extinguisher and roll-over bar, although both these were on the mods list.

The information department consisted of Winkrace Tachometer, Ammeter, Oil Temperature, Oil Pressure, Water

Temperature and Blower lb sq in gauges, all mounted on an alloy sheet dash panel.

The accelerator pedal had been positioned for ease of toe/heeling, the clutch having a very very long travel, especially so for smooth moving off. Although on the go, it turned out to be all merely a matter of dip as fast as the old gearbox could take. We were aware of a throttle stop at the end of the pedals travel, and the brakes were mighty hard. And we couldn't miss the oil warning lamp.

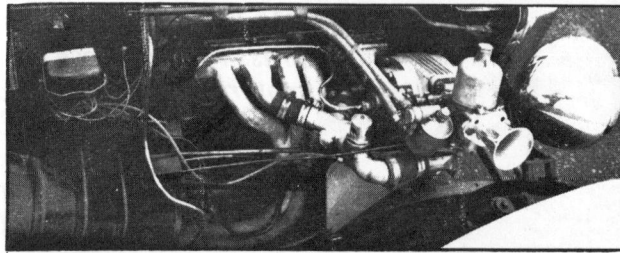

On the move, the gearbox proved to be quite noisy. It was an open car even. The throttle needed a fairly hearty blip for down changes to be made in happiness from fourth to third and from third to second.

The tiller was your actual and original plastic rimmed spoked steering wheel, complete with vicious self-centring; a feature which nearly amputated our tester's thumbs when negotiating the hairpin a little untidily.

The bucket seat proved to be a little uncomfortable being rather a tight fit. But still, whatever suits an owner and all that . . . Other intriguing individual features were the fact that one sits above wooden flooring, and 20-year-old flooring at that, as well as the T-type fact of life of a certain amount of slack in the transmission. Other features were twin fuel pumps,

twin petrol lines, Vic Derrington exhaust manifold, extra leaves as well as binding for the front springs to stiffen them up, the backs during '70 were softened to deter either rear wheel from lifting off, aero screen, twin mirrors, and an exhaust sprouting out in front of a rear wheel.

We didn't find the car at all chuckable. It needed to be treated with the utmost respect, though in all fairness, the front end felt highly sorted and nothing like an old MG at all. The handling was let down by the absence of a limited slip differential, which meant that one felt unsure through a certain amount of back-end hovering.

With the accentuated sucking of the $1\frac{3}{4}$ in SU on the end of the blower, as well as that unit's scream, there was a really goodly old row when trying. We changed gear at 6000 rpm, though 6300 was there if required. Going through the lap times achieved by the car during the '70 season makes for interesting comparisons; Club Silverstone — 1m 16s, Club Brands — 1m 6s, Castle Combe — 1m 24s, Thruxton — 1m 48.8s, and the full Mallory circuit — 1m 1.9s.

Paul's car has been trailed to Meetings and so we suppose must be regarded as an out and out Racer. But T-type competition motoring need not necessarily be this way. Fellow MGCC Member Keith Benningfield drives his example on the road, as well as Racing it.

Over 15 laps round the Stone, not once did we reckon that we took the hairpin right. We either wasted far too much time under steering across the track, lost all momentum due to lack of traction, or understeered until our driver's wrists cried enough with the sheer effort of holding the steering on.

We found cornering consisted of a series of throttle slides, albeit easily correctable. We couldn't really stick our foot down, only really enjoying the corners when they were all but over. Clear road ahead, pushing the tail out, and, looking over the edge of the bath tub, applying correction as necessary whilst receiving the full blast of the elements in one's face.

We were always aware that we were sampling a Clubman's car in the truest sense, all sports car, that is still giving of its best — and more — even after all these years. **RH-E**

R&T Happenings Dept

ALLAN GIRDLER PHOTO

M Y PARTICIPATION IN the New England "T" Register's Marathon was assured when TC owner Mike Williams telephoned for advice: Do I know anything about TC differentials?

"Swill me in a shallow pan of petrol if I don't," I said, quoting from the owner's manual for emphasis. "Working on TC differentials used to be all I ever did."

"My car goes 'click, click' when I shift."

"They all do. Nothing to worry about. Uh, it doesn't go 'clank, clank' does it?"

more than a tiny twinge. And I'm not sure there was even that.

Nostalgia wasn't a factor, then, in my enlistment for the marathon. What appealed most was a certain specious logic.

Primary sponsor is the New England "T" Register, an organization based on T-series MGs. Not a club, they say, because the members aren't people. One registers one's car. The car is the member. The owners (the owned would be just as accurate a description) tag along. The register has enjoyed a bewildering success. The co-founders expected

JOUNCING TOWARD ABINGDON

BY ALLAN GIRDLER

1000 non-stop miles in a TC, as seen (and felt) by a man who thought he knew better

"A friend of mine has a spare differential he'll lend me for the trip."

"Bring it along."

I thought I was through with all that. In 1955, I bought an MG-TC. During the 14 years that followed I raced it, rallied it, drove close to 150,000 miles on the highway, literally from coast to coast. I got frostbitten in New York and watched passengers pass out from the heat in Texas and Kansas. I replaced every part in the car at least once and learned that I am so-so at upholstery and terrible at carpentry. Enjoyable, all of it, but when the car drove away without me in 1969 I cannot say my heart was pierced by

maybe 150 owners to sign up for a newsletter, parts exchange, etc. At last count, there were 2000 MGs of TC, TD and TF designation enrolled.

Twice each year the Register hosts a get-together. The proper name is Gathering of the Faithful. The site usually is a resort hotel in New England. T-series MGs come from all over for rallies, gymkhanas, concours, parts swaps and a weekend of mutual admiration.

Parallel to the Register is the Vintage MG Car Club of Chicago. Same goals, same type of MG, a good share of dual registration.

Now. It occurred to Herb Nichols, TD owner and mem-

ABINGDON

ber of both clubs, that several other dual members drive to the Gatherings. T-series people and cars are competition minded. And there aren't many places for them to compete. Why not, he reasoned, combine all this into a marathon, a timed drive from the midwest to the GOF, a 1000-mile test, roughly (interpret that any way you wish), of man and machine. Nichols drew up some rules, secured Register support and enlisted entries. The first marathon was run in 1968, with each successive run attracting more and better-prepared cars.

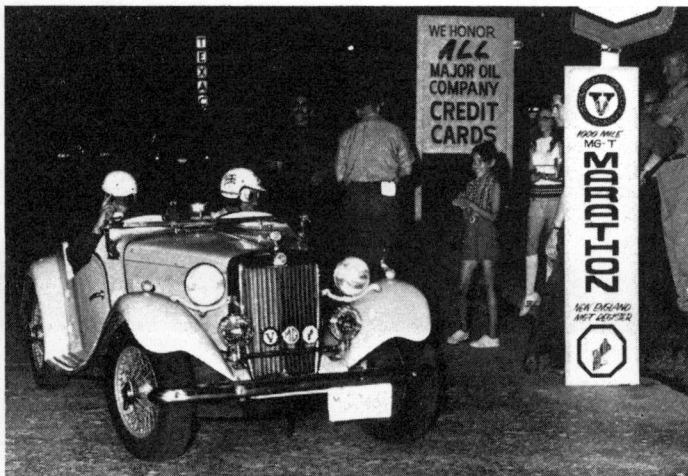

Eighteen cold, wet hours later, the windshield was still down.

The Marathon isn't a rally; the routing is flexible, the winners picked on elapsed time. Nor is it a race; most of the roads arc major highways, with limits above the practical cruising speeds of the competitors.

The Marathon is an Act of Faith, a pilgrimage to a spiritual Abingdon, to a cool little corner of New England where the year moves only from 1948 to 1954 and back again.

The object is to drive from a pre-determined point to the GOF and to do it in the least possible time, stopping only for fuel and mechanical derangements. (All former T-series owners pause here and reflect. An Act of Faith, indeed.)

There are two major rules: T-series MGs only and no non-T engine swaps. Winners are determined by elapsed time multiplied by a handicap. Scratch car is the TF 1500, in theory the T-series best suited to fast—well, relatively fast—cruising. The older the car and smaller the engine, the larger the handicap. The intent is to encourage the older cars to compete and to not lose heart when newer models whiz past.

The first marathons started from the midwest. For 1971, by popular demand, there were starting points east and west, so New England-based faithful could participate.

The man for whom I crewed didn't actually need much help. Mike Williams bought his TC in baskets and restored it, after practicing on two Austin-Healeys and an MG TF. Nor did he bring the differential. After stowing the sidecurtains, tools, gaskets, a spare half shaft, oil, gear lube, first aid kit and suitcase, there wasn't any room. Our provision for the trip was a sack of turkey sandwiches: "I asked my wife what will keep for a long time without refrigeration,

and she said turkey would keep the longest." So turkey it was.

To date, no one has actually built an MG for the marathon, but the regulars do equip their cars with the event very much in mind. Nichols' car looks like the modified TDs of 20 years ago, with 8-port aluminum head, hood strap replacing the side panels, velocity stacks in place of air cleaners, etc. Another TD was supercharged. A third has an MGA rear axle, complete with 4.3:1 ring and pinion. Doesn't climb hills very well, he said, but it surely helps on the flat. The only pure-stock TD in the western contingent did what he could and ran the distance with windshield folded to minimize wind resistance.

But the rival we feared most created terror by his very lack of preparation. Owner Bob Pickard drove his TC, Charley, 700 miles from home to starting line. Charley has 247,000 miles on the odometer, sounds like a sack of nails and holds the TC record for the marathon. A clear case of supernatural forces at work.

The western starting point was a gas station at the western edge of Michigan, adjacent to an Interstate that leads to Detroit. The route was across Michigan to Windsor, Canada, and on to Toronto and Montreal. From there we could go east in Canada and veer due south to the gathering at Waterville Valley, New Hampshire. Or an entrant could head south from Montreal and work east through Vermont.

Our TC had the lowest serial number, so we were the first car away. The start was at night, in order, I guess, to have the teams fresh during the darkness while being deluded out of fatigue by the rising sun. Which is how it works out, I found.

Mike took the first stint behind the wheel. Mind, I didn't just sit there. The copilot's duties included handing over turkey sandwiches, shining a flashlight on the coolant gauge mounted in the radiator cap and on the oil pressure gauge, located in the dashboard but minus illumination for some obscure electrical reason, and—most important—looking back to see if anything was gaining on us.

We had the idea, correct as it happened, that we were the slowest car in the western group. The entrants left at 5-minute intervals. If we were passed by the second car in one hour, we reckoned, that would mean he had 5 mph on us, and so on down the list.

An uneventful hour passed. Then another, while Mike sawed at the wheel and I peered at the gauges. Three hours out, the marathon turned into a dead heat. Charley went by, with the first TD a minute behind him, the second TD a minute after that and the supercharged TD two minutes after that. They all waved.

Very discouraging. Without consulting me, Mike and Pickard had agreed to dispense with the handicap. The TCs would face the TDs even up. We were the slowest, and all we had to look forward to was mechanical troubles for

Owner/driver Mike Williams and his triple-threat TC.

Robert Herlin's supercharged K-3 Magnette was the star of the gathering.

everybody else, a prospect not conducive to sportsmanship.

For example: All the cars were close together at the first fuel stop, save one TD which evidently had taken a wrong turn. We all laughed.

My turn. Driving a TC again felt familiar, in some ways. The tiny pedals, the upright posture, the flexed elbows were just as I remembered them. By the time I sold my TC, though, it had undergone modifications that provided half again·as big an engine and exactly twice the stock power. I had forgotten how much of the TC's wonderful exhaust was brag, and how little of that powerful roar was fact.

And the steering! Terrible! In top condition, TC steering is bad enough. With all possible modifications, it's barely tolerable. This car had a bad unit. Mike adjusted, to no avail. Either it was heavy, vague and notchy or it was heavy, vague, notchy and sloppy.

We kept Charley in sight across Detroit and through the tunnel to Canada, but then he pulled away. Past midnight, now, and getting colder. We unfolded the windshield at dusk, put up the top when it was truly dark. Mike burrowed through the parts and equipment, hauled the front side-curtains out of their bin and installed them, all on the move. Great skill involved there, although maybe you'd have to try it to appreciate it.

All my old skills came back. Stuffing the left hand into the heater, winding the right hand through the spokes of the steering wheel and stuffing it into the other heater outlet. Jamming the left leg between clutch pedal and transmission tunnel so as to unbend the knee. Folding the left leg between gearshift and seat cushion when heat from the transmission blisters the foot.

I was reminded of Le Mans—no, wait! Guiding a TC at 65 mph must require the same degree of skill and concentration as guiding a 917 at 240 mph. True, the MG gives you more time to keep between the curbs but you need a much larger arc of correction.

And at Le Mans when one's driving stint is done, one climbs out of the car, withdraws to a quiet room and lies down on a soft cot. In the Marathon the co-driver is still plugged in, bolt upright, wind chilling his kidneys, exhaust howling in his ear, the water temperature and oil pressure gauges still demanding his attention, the driver still demanding turkey sandwiches. (Actually, I just said that for effect. The demand for turkey sandwiches slackened considerably after the first dozen.)

During my tenure of ownership I did just about everything that can be done in a TC. Sleeping wasn't one of them. Mike managed, somehow, to doze while curled up with his jacket over his head. "Ya slept all night," I said later. "I had my eyes closed for 40 minutes," he countered.

Do not get an impression of gloom and discomfort. The cheeringest things happened, like passing Charley, reduced to 55 mph by (we learned later) an oil-soaked clutch. As we passed them, I waved.

Returning to acts of faith for a moment, the supercharged TD had the only service car in the marathon. Somewhere west of Toronto a fan blade broke off and sliced through the lower radiator pipe. When the back-up car arrived it towed the TD to the nearest town, where the car was left for repairs. This was the only retirement of the event, that is, the man who prepared to have the car let him down *did* have the car let him down. The moral is obvious.

We caught the blown TD before his support car did. The crew was standing next to it, with disconsolation writ large

JIM WINDMEIER PHOTO

Marathon founder Herb Nichols, right, vs a frozen thermostat.

on their faces, allowing me to use that famous racing quip: "Nothing trivial, I hope."

Mike laughed and laughed. There's a fatigue factor here. Canadian winters buckle the pavement something fierce, and a TC goes over the buckles in great crashing leaps and bounds. So do the occupants. My image of us sailing topward in unison reminded me of my favorite Laurel and Hardy movie. I said as much. When we stopped to adjust the water pipe to the heater, Mike climbed out with, "Be right back, Ollie," and I laughed. It seemed funny at the time.

As the sky lightened, our spirits rose. At the 10-hour

ABINGDON

mark, we had covered exactly 570 miles. The TC course record was a 53-mph average, so we looked like contenders again. Especially when we passed the stock TD, hood open, occupants puzzled, just outside Toronto.

Another high point: A Canadian T-series fan provides coffee and doughnuts for the Marathoners. This year he was invited to crew a trans-Canada rally, which he did, leaving his wife to brew gallons of coffee, buy sacks of doughnuts and load them and three small children into the family station wagon at dawn and spend several hours ministering to passing Marathoners. It was indeed a welcome wagon. I am only sorry that several entrants weren't willing to use five minutes taking advantage of her hospitality. But the kids said they enjoyed eating one dozen doughnuts each.

Somewhere between Toronto and Montreal—I know, that's a long somewhere, but I was driving, not navigating—we caught one of the eastern contingent's TDs. They started half an hour before we did, so we were ahead, and craftily tucked in behind, as they had driven through Montreal previously. A good thing, too, because our map reading would have got us lost.

Outside Montreal we passed the last mandatory gate, a tollbridge. Here I made our Maximum Blunder. We had the choice of going east in Canada then veering due south to the center of New Hampshire, or going south from Montreal into Vermont and then east.

Mike had worked it out on the map, and wanted to go east then south. The TD headed south.

"Better roads the other way," Mike said.

"Those people have proved they know better than our maps," I said. My will prevailed, mostly because I was driving at the time, and I quickly wished it hadn't.

The land between Montreal and the ocean is covered with mountains and valleys, the latter running north-south. We got south quickly but then learned we had to meander over the hills, on narrow country roads.

Any other occasion, fine, but time was running out. Strange, the human mind. While we crawled across the endless plains, all we could think of was the machinery. For the last 100 miles the gauges, the pressures and temperatures, the odd noises were all forgotten. We had more important things to worry about. The 57-mph average fell victim to 40-mph corners.

We couldn't find the instructions in the jumbled equipment. We took a wrong turn. "Haw, haw," said a local with that rural good humor that makes big cities so attractive, "all you MG guys are a-gittin' lost."

Indeed. We couldn't even find the finish line. 'Round and 'round the hotel parking lot we whirled in a shower of mud and curses, furiously beeping the horn in the 3-long pattern supposed to notify the timers that we had, at last, arrived.

After one or two circuits, we followed the waving hands to a table containing the scorers, who collected our toll tickets and clocked us in. The last 7 miles took 17 minutes. The finish line was hubcap deep in T-series cars. Not realizing that only a fraction of the Register cars run in the marathon I thought we were among the final finishers. Not content to compare our efforts with Le Mans, I glumly thought of Indy, another place where Rookies Never Win.

There is a happy ending but first, the Gathering. What the Faithful mostly do when gathered is talk about and look at their own and each others cars. There was a rally, a concours d'elegance, a swap meet, a gymkhana and a raffle. All low in pressure. We missed the rally but I inferred that it was chiefly a drive around the New Hampshire countryside. Beautiful place to drive, with or without checkpoints. The concours had several spectacular cars, a supercharged K3 Magnette for example, and a couple of incredible restorations, but many entries were simply daily MGs, washed and waxed for the occasion. The gymkhana was more gimmick than go-fast, with a lap in forward gears and a lap in reverse. And the raffle prize was a J2 Midget in dreadful condition; the sort of prize you'd be happy to win and happy not to win.

Mike struck a blow against specialization. His TC placed fourth in class in the concours and was both the only marathon car to win a prize and the only prize winner to run the marathon. Next, he had second-best time of day in the gymkhana. I think there should have been some sort of Big Man of Gathering award, so he could have won it.

On to the banquet. Very clubby, as one would expect and enjoy if one likes MGs, as I do. There were door prizes, one of which was a book, courtesy of R&T. That won me many smiles and appreciations, although in fact I hadn't known about our donation until it was made.

Then they began to announce the marathon winners. Consternation reigned at the marathon table when Bob Pickard won fifth place. Remember, both TCs agreed at the start to waive their handicap. But Bob was eighth on raw time and fifth in the awards, which we took to mean that the waiver hadn't been accepted.

Sure enough, the announcers worked their way higher and higher in the placings, reeling off the names of teams behind us on raw time. Then, the grand prize, first place, let's hear it for Mike and me!

Mike just sat there, half pleased to win and half upset because he had been given the win against his intentions. I loudly whispered that he'd better get up and accept the prize or he'd hurt the feelings of the men who had done all the scoring and set up the rules. His good manners prevailed and he agreed to be the winner.

The next morning Mike started his second 1000 miles of the weekend. Solo. Pleading the press of business, I hitched a ride to Boston in a TF. Along the way we stopped at a garage near the hotel, to deliver parts and advice to a TD that suffered a broken axle in the slalom. That's the way these things go sometimes, he said. Everything went right for me, so I enjoyed the gathering. So did those for whom everything went wrong. I don't know a more convincing recommendation for an event than that. ⊙

Well-wishers and just plain curious thronged the starting line.

The winner, at speed and in reverse.

MIKE WILLIAMS PHOTO

1949 MG TC

CONTINUED FROM PAGE 75

folded flat (in which condition it will go three or four mph faster).

Acceleration is sedate—a little slower than the current Austin America, and markedly slower than latter-day Midgets and Sprites. This is to be expected with an output of 54 bhp from a 1250-cc pushrod four. With its long stroke (90 mm) and a stumpy axle ratio of 5.12:1, the engine pulls well in the lower gears. This ability, along with an effective fly-off handbrake, means that stopping and restarting on steep hills holds no terrors. But the hydraulic drum brakes—little 9-inchers within the huge 19-inch wheels—aren't very powerful, even for the moderate performance and relatively low weight of the car. And lacking any servo assist, they require a pedal effort that makes following closely and charging stop signs decidedly unattractive. But then, we wouldn't want to do this in any car of the period; brakes have come a long way in all cars since then. Fade is also a problem on the TC, which ran out of brakes before completing the usual six successive stops from 60 mph.

In the comfort and convenience department, the TC has individually adjustable seat cushions sharing a common, adjustable backrest. Legroom is good for a tall driver, but a shorter one must move right up on the already close steering wheel. The doors are hinged at the rear, permitting one to back into the seat, a maneuver hampered a bit by the short doors and the lack of thigh room between the seat cushion and the big steering wheel. The upholstery is attractively simple, with a nice choice of colors keyed to the exterior paint—green with cream (our test car), tan with red, etc. The grille uprights are painted the same color as the interior, enhancing the contrast. And of course there is a badge bar. Rotating map lights-cum-switches are mounted next to each main instrument, map pockets are sewn

into each door panel, and a very useful passenger grab handle is fastened to the left-hand side of the fascia. The operation of the wipers, with windshield-mounted motor, is minimal. Physical weather protection is reasonably effective, if time-consuming to erect; the fabric top leaves generous headroom (39 inches from seat to top). The side curtains, which include rear quarter windows, fit into sockets in the door tops. Behind the seats is a luggage platform which will accept one medium-sized and one small suitcase. Behind this, "outside" the body, is a 16.2-gallon fuel tank with quick-release filler cap. And behind that, finishing off the architectural exercise, is the exposed spare tire.

Outmoded design practices aside, the MG TC is a better built car than most of the machinery that has left England since. The reputation of British quality, which has taken such a beating in recent years, was high in the immediate postwar period and the TC justified it. Both the charm and the quality of the TC are responsible for the relatively large number of TCs that are still lovingly cared for, both in the British Isles and the United States. Concours examples command one and a half times the original $1895, with even tatty ones being worth that. The far-more-numerous TD is just beginning to come into its own now that the ravages of time have depleted the ranks of good ones, but the later model, for all its clearly superior performance and handling, will never deserve the term classic. The difference, of course, is esthetic. When the MG designers reduced the height of the wheels and fenders and moved the radiator, engine and cockpit forward, the resulting TD lacked the grace and dignity of its predecessor.

Our test car is the property of Elaine Bond, who had owned a TC in its heyday and a few years ago was given this one as a Mothers Day present. The present TC normally lives in the reception area of our office building, but was made ready for testing in an afternoon. The only complaint the TC made after going back on the road was to boil away some of its water on extended runs—certainly a display of temperament permitted a dignified veteran among all the modern upstarts.

1946 MG/TC on the road

Probably the first post-war car to have "classic appeal" and particularly in America was the MG TC. It had all the vintage virtues in its appearance with running gear which could still be obtained; even when obsolete it could still be run, maintained or restored at minimal cost. This is no longer quite true but there are a number of MG specialists around who are producing body parts to original design and you are unlikely to be stuck for long if you are restoring one.

This particular one is chassis No. 1601 built in 1946; the present owner Mr. Fuller bought it in 1969 ostensibly for his daughter, but by the time he had driven it back from Cambridge and discovered that it had bent wheels and wouldn't steer he had to look deeper. Having then seen that the car had been recently rolled, he had to embark on a major strip and rebuild and his daughter had to look elsewhere. Three years of summer evenings later it emerged in this pristine condition with just about everything renewed or at least refurbished—new chassis, new axle casing, new wood frame, new panels and reconditioned engine and gearbox. It must now be better than original and it has certainly proved its appeal with the number of concours prizes already achieved.

The road test below is from The Autocar of October 17, 1947 and contains delightfully period touches. The figures about which they enthused would bring a wry smile even to the owner of a standard 850 Mini which returns much

the same—maximum 75 mph, 0-30 mph 5·7 sec., 0-50 mph 14·7 sec., 0-60 mph 22·7 sec. The price then including purchase tax was £527 16s. 8d. One that is "ripe for restoration" could cost as little as that now; this one would be a lot more.

In a motoring world in which there is so much talk as there is today of rationalisation, and in which cars tend more and more to resemble one another in appearance as well as in performance for a given size, the MG Midget two-seater stands unique. Yet an interesting point, as shown by recent public utterances on export subjects as well as by other sources of information, is that this car does not appeal only to the trials-minded and youthful fraternity of motorists in this country. On the contrary, it is gaining more and more of a following in other countries, including the USA, and has reached a position where it can be regarded as one of our more exportable cars in terms of proportion of total output of the model.

Today it is certainly a class alone among cars made anywhere in the world as a sporting type retaining the conventional outward appearance of the "real" car dear to the hearts of enthusiasts in years gone by—that is, by displaying its radiator, or at all events a normal grille, and lamps, and in not having gone "all streamlined". It is a model, too, which more than most cars has evolved through the years, with its beginnings in that much smaller Midget of seventeen years or so ago that instantly registered a success. No car has done so much to maintain open-air motoring and to support the demand that exists all over the world for sports car performance and characteristics in a car of not exorbitant first cost and at moderate running costs.

It offers a great deal in sheer performance, yet is not just a sports car with an appeal limited to special occasions; instead it is in every way a perfectly practicable car for all occasions where

As it might have left the factory in 1946 GAD 518 is in superb condition and has taken three years from being just a runner

two seats are sufficient and the fresh air style of progress is preferred. Actually, the all-weather equipment is good, the hood being easily erected and the side screens likewise, and they turn this car into a very reasonable imitation of a permanently closed car for bad weather use

The Midget is in no way more difficult to drive than the ordinary family saloon, but given the type of driver who usually falls for such a machine—not necessarily a youngster—and who likes to use the gearbox, the performance becomes quite vivid. That is not to suggest that the gearbox has to be used in the manner of a pump handle whether the driver chooses or not; the 11 hp rated engine that the TC Midget possesses has quite a range of flexibility on top gear, and the car is tractable in traffic. On the other hand, with an engine that will rev very freely without complaint much more can be made of the performance, of course, by using indirects that offer maxima as high as 60 mph on third and 40 on second.

Owing to the handy size of this car, its ability to pass safely where a bigger car would be held back, and the way in which it regains its cruising rate after it has been checked by other traffic, the Midget is almost as fast a car, over British roads, as can be found today. One feels, too, from its ability to take hard treatment and to hold speeds between 60 and 70 mph, apparently for as long as roads in this country permit such motoring, that stretches of motor road offering far more opportunity of sustained speed than ever is found in this island would not "melt" a Midget engine.

The handiness of the car, the way in which it helps the driver in its manner of cornering, its "quick" steering, are big factors in giving it unusual average capabilities without an extremely high maximum speed being attained. The present car has been handled over a considerable distance in conditions which provided crowded roads, and also over routes on which traffic had been thinned by seasonal and petrol considerations. In both circumstances the average speeds were exceptional, a 40-miles-in-the-hour showing

seeming always to be within its reach on a journey of any length, while, when roads are clear, figures such as 44 and 46 mph averages have been obtained. When the car was being timed by The Autocar's electrical speedometer to be travelling at 75 mph, the car's speedometer showed only 73, an unusual state of affairs. In other more helpful road conditions subsequently the car's speedometer was seen at the 75 mark.

Sense of Accurate Control

Always one has the feeling of being able to make a fast run easily in the Midget, for it responds so readily to all the controls and is so quick—eager, it seems—to get moving. The biggest factor in this and other directions, apart from the actual performance available, is the complete sense of command which the driver feels he has over the car at all times, including the major features of

brakes, steering and road holding on corners. The Lockheed brakes deserve special mention, for they deal most effectively with high-speed braking, and also are powerfully smooth in low speed applications.

Merits and demerits of normal versus independent suspension can be urged in the main to the latter's marked advantage, but there is no doubt of one fact in this connection. The normally sprung car, rather hard sprung, as in this instance, does let the driver gauge within close limits the speeds at which he can corner safely fast. After a little experience of it one finds oneself holding quite high speeds round bends in the Midget, and the car steering to a close course only a foot or two out from the near side verge. Such a half-elliptic suspension has, of course, the counter-balancing feature that it is on the harsh side over

poor surfaces, but on the Midget this tendency is by no means excessive.

It is a trim and appealing little car in its general arrangement and very practically laid out, besides offering a considerably higher accessibility factor than is usual today. One quickly comes to feel an affection for its efficiency and willingness, and in all respects, including performance, it is "man-size", with no suggestion of the tiny car about it.

Driving Position and Controls

Doubly important in a car of this type is the driving position. The Midget is provided with an adjustment for the seat back rest, which is in one piece, although there are two separate cushions, whilst also the spring-spoked steering wheel is telescopically adjustable and can be placed ideally for full power of control. A feature much appreciated or disliked, according to the point of view, is the fly-off type of handbrake lever—in The Autocar's view a form of control to be highly commended for its certainty and positiveness of operation. A more comfortable position for the left foot off the clutch pedal would be welcomed.

The gear change has synchromesh on second, third and top, and with a short vertical lever, which is well placed, this works very well for really quick upward and downward changes when the utmost is being made of the performance potential. The instruments include a rev counter, and the engine can be taken round to 5,500 rpm with celerity, and it will readily go beyond that figure.

One does not think of this car in the usual way in terms of top gear climbing ability. Actually, however, the capabilities in this direction are good, for the power-to-weight ratio is favourable, but it is a delight to drop to third and fly over the gradients that bring the speed down at all appreciably on top. As to steeper gradients, second gear lets the car tear up a hill of 1 in 6½ calibre.

The head lamps are good for fast night driving. Starting from cold is immediate, and not much use of the mixture control for the twin S.U. carburettors is needed before the engine will pull properly. An excellent point, of value here, but still more so in territories where filling facilities are widely spaced, is the big petrol tank, giving a range of action of approximately 400 miles.●

A NEW AWAKENING

by Jim Gilbert

The introduction of the MG-TC 25 years ago created a new automotive genre and alerted drivers to a new criteria of vehicle performance.

Maybe British Leyland says it best in their current ads: "When the MG-TC first came to America, it was love at first sight."

The consensus is it was more than love. It was the start of countless businesses involving "sports cars" and their accessories. Cultism grew rapidly, the ticket to ride punched with a badge bar. How many millionaires were made by the sudden post-war infatuation with sports cars? How many fortunes lost? How many stories begin: "I had a chance to be the first Jaguar dealer in southern California, but . . . " Or: "So I mortgaged the house, borrowed ten-thou from my wife's parents and started importing the HRG, the Jowett-Jupiter and the Singer . . . "

But for most people the MG-TC was, indeed, the start of a love affair. It *was* a new awakening, the introduction of a new automotive genre. Drivers were alerted to a new criteria of vehicle performance. Alternatives to the cattle cars masquerading as Detroit sedans were suddenly available — for a reasonable price.

And the car that started it all — whatever it was — is generally accepted to be the MG-TC. Nearly every performance nut has a story about the TC, and most of these early owners are probably kicking themselves for not holding onto the TC as a collector's item.

Today, nearly a quarter-century after its introduction on the market, the TC still holds its "modern classic" appeal. It's tall and thin with long, sweeping fenders. It's underpowered with a hard ride and sloppy steering. But, hey! It's a *classic*, man!

Your basic TC, you understand, has a 94-in. wheelbase with a 45-in. tread, front and rear. The overall length is 139-in., width 56-in., and it sits 53-in. A two-seater, of course, with room for three *Road Test* Magazines in the back, the TC has a 1250cc (76.3 cu in.) four-cylinder which produces 54 horsepower at 5200 rpm. Top speed is somewhere between 75-80 mph.

With a four-speed, the 0-60 run takes some 21 seconds, about the same reading for a quarter-mile dash at the drags. The TC, as easily seen, was no powerhouse of performance, but it had class, *elan* and succeeded in shaking the cobwebs out of a generation of complacent drivers.

Mitch Leland is a creative planner/builder/developer who works out of an historic estate dating from the "Lucky" Baldwin days of California. He's been involved with a variety of modified cars, and the MG-TC featured here is (or was) his latest — was, because he just sold it to a wealthy Chicagoan who popped $5,500 for it and had it flown back on a chartered plane. (Mitch's current project is a 300SL, which you will see in *Road Test* upon completion.)

Mitch is an automotive perfectionist. When he found his TC, another enthusiast had started rebuilding it but without the eye to detail Leland demands. It wasn't a basket case, but it soon became one. Leland stripped it totally and started over.

Restorations are an expensive, time-consuming business. Perfection in this field is demanding with pride of accomplishment the only reward. The following gives an idea of what Leland found necessary before the TC was finished.

The frame, being the backbone of the car, is first. It was completely straightened and aligned, and all rivets were tightened. Cracks had appeared at the front spring hangers and these were welded. The frame was sandblasted and repainted. A new set of rubber bushings was added, and front and rear springs were rebuilt.

The shocks were rebuilt with SAE seals, something that had never been done before, so Mitch takes credit for originating this operation. The shock arms, brake drums and front axle were ground and painted.

TC steering is a joke. The car would wander all over the road, bouncing off white lane lines like a billiard ball on a bank shot. Mitch did his best to fix it. He obtained a special heavy wall tube and column, and a Tompkins steering kit. A higher quality roller bearing set was installed in the steering box, and a bronze bushing was added to the steering column top. A new Pitman arm and sector shaft were added, the shaft hard chromed. This all worked to tighten the steering and make the car respond better to the driver's intentions. The old "goes where it's pointed" never applied to the TC, but at least it doesn't bounce off walls anymore either.

The brakes on the TC contributed to its appeal, but they weren't exactly 747 anchors. Mitch found his in horrible shape and totally redid them. The lines are all

steel — not copper, mind you — with double flared ends. Hoses and shoes are new and the wheel cylinders are rebuilt. The master cylinder was machined from .316 stainless. The system itself is a vacuum booster type and Mitch has no complaints with it.

Probably the first eye-grabber on a TC would be the giant 19-inch wheels — *wire* wheels — great circles of spokes. All of them were straightened and spokes replaced where necessary. The wheels were sandblasted and chromed, and new bearings installed. A set of hard-to-find Dunlop tires were found (anyway) and a set of innertubes (remember them?) from Montgomery-Ward are used. They're not natural rubber.

The trans and rear end didn't cause too much trouble. Both are rebuilt and have new parts and bearings where necessary to bring them up to snuff. The universal has a new SAE grease seal, and the drive shaft was balanced.

The engine received extensive work. The three-bearing crank is .010-in. under, magna fluxed and micro-polished. The rods are shot peened and polished.

The exhaust manifold was porcelainized, a very rare racing type valve cover was was surfaced, and the exterior ground and filled. New freeze plugs were added. Full skirt aluminum pistons, .020 overbore, are installed along with a stock TC camshaft.

The head was ported and .100-in. ground off. Bigger valves and new guides were added, but the springs weren't replaced. Mitch notes that there is some float over 4700 rpm, but that could be due to the long duration of the TC cam.

The exhaust manifold was porcelized, a very rare racing type valve cover was installed, and the tappet cover was specially handmade to accommodate the vent pipe. A late model TF-five-bladed impeller water pump was added.

The carburetors were Teflon sprayed. It won't come off and gas won't stain it. However, to apply the Teflon, the carbs were heated and this caused some problem in reassembly. This causes some problems in idle. The air cleaners are original and very difficult to obtain.

The engine is completely balanced right down to the water pump for smooth operation. There is a TF oil filter and pump.

The motor mounts have new rubber. The front is of 3/16-in. steel instead of the 1/8-in. original type, and the rear one is aluminum machined with a SAE seal.

Hooking it to the trans is a new clutch and pressure plate.

The body was completely stripped, eventually repainted in lacquer. Mitch fitted galvanized sheet metal behind the gas tank and the inside side curtain box. The hood is specially reinforced at the corners to eliminate the usual cracks at these points. Both the tool and the battery boxes have heavy duty bottoms. The battery box can be removed by undoing the brass screws holding the i.d. plates. This permits the 30 amp battery to be removed without taking off the hood.

New floorboards of .028 aluminum sheet metal have been laminated to the underside.

The doors have special turnbuckles inside for alignment and better support. All corners have sheet metal reinforcements. All the hinge bolts are stainless and the bottom ones go through to the rear fender for added strength. The latches are brand new and, as Mitch puts it: "impossible to get," so how he got them we don't know.

The lights are a source of pride with Mitch. The heads are Lucas tri-pods, as is the rare road light. The directional lights are the most desirable for TCs. Mitch handmade them since originals are unobtainable. They also serve as brake lights.

The dash and interior are completely redone. All instruments are rebuilt and new tach and speedo drives were installed. There's a special directional light buzzer and an added auxiliary fuel pump switch. A new version of the original steering wheel was found and the pedals are black chrome. A new rubber tunnel is used. All the wood in the seats is new and treated with preserver. The leather upholstery is English Conolly Bros.

Probably the last major change is the gas tank. It is pretty common for TC tanks to leak. Mitch decided his wouldn't. It has a new 16 gauge bottom with a plastic liner inside. The sending unit is specially machined with an O ring and stainless machine screws fitted with O rings under the heads. This effectively stops the dribbles.

There is not a single part on the TC which has not been apart. Everything has been rebuilt, painted, chromed, etc.

Mitch's TC has won more than its share of Best in Shows and class trophies, but it isn't exactly pampered. With all the work, Mitch drives it whenever the desire strikes. However, he feels the sting of every little scratch, the bruise of the tiniest nick. When he was sufficiently tired of bathing the undercarriage after every drive, he accepted the offer from the monied Midwesterner.

But the TC still lives, as it did 25 years ago. Rock-firm ride, wind blowing through and attacking the kidneys, there's still that *something* that made the TC an automotive milestone. To every man it's something different, but for most it's just plain fun.

Is that what the TC did? Bring fun to driving? If that's true, one of these days we all may well ask: "Where is the TC now that we need it?" ●

Anatomy of a car
The 1946 MG TC

We look under the skin of the most popular of the early perpendicular Midgets, bristling with the wire wheels and slab tank character

Enthusiasm for the off-shoots of the Morris Garages is probably unequalled on a world-wide basis; other one-make clubs can doubtless master similar enthusiasm on a more localised basis but love of MGs is worldwide. Perhaps the proximity of so many likeminded individuals exerts a blinker effect because criticism of anything MG renders the critic liable to banishment to Coventry where other cars are made. However no-one would deny the classic appeal of the T-series MG, just one of a long series of Midgets designed to give sporting motoring for low cost; at rest they are undeniably good-looking with the lines of the pre-war period cut to a perfection that many tried to emulate; on the road their performance could only be described as adequate in postvintage thoroughbred terms but in performance per £ they were remarkable value.

The MG TC came into production in November 1945 at a cost of £480 including £105 purchase tax. They were the logical development of the original T-series, subsequently called TA, via the short-lived TB, a 1939 model sunk by the war. When the TA was announced in July 1936 it showed a marked improvement over the replaced P-series; a larger engine

more than offset the weight increase and the *Light Car* declared that the TA was the fastest MG that it had tested with a standing quarter mile in 21.6 sec and a maximum in one direction just over 80 mph.

Came the TB and the engine dimensions were adjusted from the long-stroke 63.5 x 102 mm., 1292 cc to a more revvable 66.5 x 90 mm., 1250 cc;—both with push-rod operated overhead valves; a counterbalanced crankshaft with steel-backed bearings gave greater smoothness at the higher revs—peak power at 5250 rpm rather than 4800 rpm—and the final drive ratio was dropped from 4.875 to 5.2:1; such a change only raised the revs at 60 mph from 3600 rpm to 3800 rpm but improved the acceleration.

The TC was its direct descendant with very few changes but what modifications there were, showed that some thought had gone into the car during the war years. The new Midget was some four inches wider which allowed bigger, more comfortable seats and more elbow room. The channel chassis remained the same but the semi-elliptic leaf springs used all round were shackled at their rear ends as opposed to the sliding trunnions of the TB, and the shackles

incorporated rubber bushes which gave im proved road shock absorption; Girling Luva hydraulic dampers were also adopted. Othe changes were mostly confined to cleaning u the design; under the bonnet, a new large toolbox was accessible from either side behin the scuttle-mounted 12-volt battery in its ow accessible box. The wiring was also improve and better ordered than before.

So much for the TC development. The basi design was simplicity itself with none of the expensive sophistication that other designer of more costly cars had adopted, not even a overhead camshaft design that MG had droppe pre-war at the introduction of the T-series. The chassis uses channel section of varying dept which runs straight from dumb-iron back to the rear of the petrol tank, over the front axle an under the rear. Such frames are noted for thei floppiness unless well-braced or unless the are particularly deep; the TC frame is shallow but has cross members effectively at front anc rear of each semi-elliptic spring and around the scuttle. Tie-bars at each end of the chassis only add a little to rigidity, but front engine

Mr. Fuller's recently restored TC with everything as it was in 1946; the proud owner can be seen in the side mirror. Under the bonnet the perfection of restoration is shown; Note the simple cast manifold complete with octagon

MG TC DATA (1947)
PRICES (Including car tax) £527.67
ENGINE
Capacity : 1250cc (76.3 cu in)
Configuration : In-line four with pushrod ohv
Bore x stroke : 66.5 x 90mm   (2.61 x 3.54 in)
Compression ratio : 7·4 :1
Carburetters : Two SU
Power : 54.4 bhp at 5200 rpm
Torque : 64 lb.ft. at 2600 rpm
TRANSMISSION

Gear ratios :		
Top	1.00	
3rd	1·35	
2nd	1·95	
1st	3·38	

Final Drive : 5·125 : 1
Mph per 1,000 rpm : Top 15·5
RUNNING GEAR
Suspension : Dead beam front axle located by semielliptic leaf springs. Live rear axle located by semi-. elliptic leaf springs.
Dampers : Girling-Luvax lever-arm
Brakes : Lockheed drum brakes hydraulically operated 9in dia.
CAPACITIES
Fuel tank   13½ gallons
Oil   10½ pints
Water   14 pints
DIMENSIONS
Wheelbase :   7ft 10 ins.
Track :   3ft 9ins.
Overall length :   11ft 7½ins.
Height :   4ft 5 ins.
Width :   4ft 8 ins.
Weight :   16½ cwt.
PERFORMANCE :

In gears	Top	3rd
Mph		
20-40	12·2 sec	8·0 sec
30-50	13·0	9·3
40-60	17·2	11·6
Standing starts		
0-30	5·8 sec	
40	9·0	
50	13·9	
60	21·1	
Maximum speeds		
Top	73·2 mph	
Third	61	
Second } at 5300 rpm	42	
First }	24	

Brakes (from 30mph in neutral)

Pedal load (lb.)	Retardation	Stopping distance
25	0·48g	63ft
50	0·75g	40
90	0·85	35

FUEL CONSUMPTION
(at steady speeds in top)

30 mph	43·5 mpg
40	39
50	34
60	28·5

Overall fuel consumption over 372 miles 33 mpg (8.6 litres/100km)
STEERING
Turning between kerbs :   37 ft.
Turns lock to lock :   1½

Figures recorded by Motor for issue 2nd April 1947

Superbly restored MG TC; this one is a regular concours car that is driven to and from meetings by its owner/restorer Mr. Fuller. After three years of summer evenings it emerged in this pristine condition, new chassis, new axle casing, new wood frame, new panels and reconditioned engine and gearbox

1946 MG TC

Anatomy of a car ...

The front axle is a typical beam and sits on the front springs; a worthy fetish of the era was a greater respect for steering geometry than exists at present. Thus the 7° king-pin inclination and 3° positive camber are set to give very nearly centre-point steering which keeps the steering light enough to allow a useful contribution to stability with high castor angles—8° initially but reduced for weaker TC owners in 1948 to 5½°. In fairness 4.50 x 19 tyres do not have the high natural trail or castor that modern radials do.

Rack and pinion steering hadn't really arrived by then for the masses and the TC used

Classic cutaway of a classic sports car, one of a long-line of Midgets and the last to use a beam front axle. See Classic Car October 1974 for Road test and 1946/49 MG TC Mini Manual

mounts and the boxed member joining, or keeping apart, the rear spring from mountings add usefully; the major contribution comes from the pressed steel scuttle and the cross member which carries the rear of the engine/gearbox unit.

Flat rear leaf springs are outset from the frame to give a wide spring base—better for roll stiffness without excessive bump stiffness. However at the front the spring base has to be kept down to give adequate steering lock (still not good at 37ft) while the springs are shorter (and stiffer) than at the rear. The *Motor's* period description declared. . . 'The springs as a whole are provided with limited motion, and the practical results on the road are a faithful interpretation of the theoretical anticipations with this type of layout' which means that it should have been stiff and jerky and it was.

Rubber as a shock absorbing medium wasn't altogether trusted in that era for its possible effect on steering accuracy, so the front spring front bolts ran direct in the eyes but the front of the rear spring used a Silentbloc bush; both sets of shackles used loose rubber bushes. Both front and rear springs were set to be flat when laden which improves the lateral location.

a Bishop Cam and Gear steering box with transverse drag link to the nearside and a full-width track rod, with ball-joints at both ends of each to absorb wheel shocks. Shock absorber and rear axle design seem to have changed remarkably little over the past 28 years, as the TC used lever arm hydraulic dampers albeit with less developed innards than today's designs, but the rear axle design complete with drum brakes is remarkably 'modern'.

A one-piece large diameter prop. shaft leads into the back of the 4-speed gearbox; the use of synchromesh on the upper three ratios marked an improvement. The synchro wasn't particularly powerful but the gearbox had ample capacity for more powerful engines and with its compact dimensions was a popular replacement for cars other than MGs. *Motor's* comments reflected the era. 'This (synchromesh) is a comparatively new arrangement for the T-series and one which makes possible very

slick gearchanging without double clutching. Going down, the procedure is to hold the throttle open and steadily press the lever towards the desired lower gear position, while changing up can be done with great rapidity. Against this, quiet changes without the clutch require a much higher than average degree of skill, if they are to be brought off successfully. A Borg and Beck single plate clutch is bolted to the back of a flywheel that was much lighter than on the TA.

The engine too follows a 'modern' pattern which has become standard British Leyland concept, although crankshafts now boast five main bearings for four cylinders. What is interesting is that the output of 55 bhp from 125

or 44 bhp/litre is still up to modern mass-
production. This is achieved with push-rod
operated overhead valves which are slightly
inclined in a mildly wedged bath-tub combustion
chamber of 7¼-7½:1 compression ratio. By this
time the original philosophy of 'get the exhaust
gases out well and the inlets can look after
themselves' had largely disappeared. Obviously
it was still advantageous to retain separate
exhaust ports for cooling purposes but the
inlet valves at 33 mm were 2 mm bigger than

the exhausts; siamesed inlet manifolds were fed
by twin 1¼-in SU carburetters of the semi-
downdraught type, Valve timing gave 35° of
overlap with 248° inlet dwell and 256°
exhaust. The exhaust system wasn't particularly
inspiring with a simple cast four-branch manifold leading into a single downpipe and back to
the Burgess silencer under the driver's seat, with
flexible piping into the silencer to absorb engine
movements from the rubber mountings. The
power output was good for the era and the
torque curve was flat peaking with 64 lb.ft. at
2600 rpm, but it stayed over 58 lb.ft. from 1300
to 4800 rpm.

The cast-iron block/crankcase is split on the
crankshaft centre line with a deep cast alloy wet
sump underneath; the gear-type oil pump is
driven from the chain-driven cam shaft-via a
skew gear and feeds through a full-flow filter.
Three-ring pistons run in dry liners and there
are cooling water passages between each cylinder, water circulation being via a water pump;
fan and thermostat were fitted.

The TC is very much a two-seater but there
is some luggage space behind the front seats;
separate adjustable squabs were supplied but
the backrest was in one piece and could be

adjusted for rake. The batteries used to be
carried under the luggage area but the single
12-volt unit was carried on the TC's scuttle in
its own box. Classically placed at the back, the
slab tank with its quick-action filler takes 13
gallons, enough for over 300 miles.

In absolute terms the performance was
nothing very startling with 0-50 mph in 13.9
seconds while the *Motor* only recorded a
maximum speed of 72.9 mph with the hood and
sidescreens on and 73.2 mph with the screen
flat. But in 1947 for just under £530 it was
about the best value on the market; it was
smart, went faster than the popular saloons and
cornered well on most surfaces.

Motor declared that 'the MG is patently
designed to be driven faster than most and is
sprung accordingly. Consequently it must be
admitted that at speeds below 40 mph, except
on roads with impeccable surfaces, the ride
becomes increasingly harsh as speed diminishes,
and this effect is enhanced by the slightness of
the upholstery'. The testers finish: 'To sum up,
the TC-type MG provides a very adequate means
of transport for those who place performance
and stability and an ability to go almost anywhere, high on their list of requirements.
Mechanical accessiblity is above average, and,
by present-day standards, the car represents
very good value at the all-in price of
£527 16s 8d.' ●

JUST REMINISCING

Ian Fraser takes to memory lane by MG TC

PHIL SAYER

JUST REMINISCING

THREE YEARS MY SENIOR, HE WAS TALL AND thin, more man than youth, and we had in common a dislike of sport coupled to an abysmal academic performance. These disqualifiers were the real reason why he could not bring his car into the school grounds; had he achieved prefect stature it would have been permissible to park beneath the clock tower in the area reserved for those in positions of power and privilege.

Every morning on the way to school, I deliberately walked up the street where the car was parked, nose first into the kerb, just far enough from the gates to be invisible to those in authority who could have stopped him, as a boarder, from having anything as outrageously independent as a car. Of course, everyone *knew* he had a car, and he knew that everyone knew, but he was mature enough not to flaunt it. Unable to contain myself, I eventually swallowed hard and asked him to take me for a ride. He thought for a few moments, then said he would but only around the block because he wanted a different parking place, anyway. Conspiratorially, we slunk out of the grounds one lunchtime and along to where the car was parked. It was autumn and cool and the wind brought showers of brown leaves down from the poplar trees so that they blocked up doorways and rustled along the road. The car was half buried in leaves. We carefully brushed them off the long bonnet and the taut canvas roof so that the crunchy piles around the spindly wire wheels increased in depth. Finally we were ready. The deep red paint, the beige leather trim, the 19inch wire wheels, the cutaway doors with their side curtains in place. My host coiled himself up with a strange technique to ensure that his legs, which were almost as long as the car, were aboard, whilst I awkwardly grunted my way down into that inviting seat. The door notched closed and I was sitting not only in my first sports car but also in my first MG. It would not be the last.

Ignition on, fuel pump ticking away. He pulled out the choke, then the starter. At once the little pushrod 1250cc engine burbled cheerfully into life. As agreed, we went around the block to another parking place, probably didn't go above 25mph and certainly never achieved the dizzy heights of top gear. It didn't matter: I was hooked and there was no turning back. Yet paradoxically I never did own a TC. Well not all of one, anyway.

Trouble was that by the time I left skool and entered what could laughingly be called gainful employment, the TC had been replaced by the TD; and by the time I had fallen in with people of like mind, the Mk2 TD was a reality and the TF was imminent. But there were some who firmly believed in the TC, regarding the independent front suspension of later types as a novelty that really would not catch on for sports cars. There were also those who, slightly less impoverished than I, chose TCs for straight financial reasons: they could not afford anything more desirable, even though their fancies ran to XK120s and beyond.

Posing as a newspaper reporter, I contrived to get myself invited along on a two-day car trial. It was there that I met my second TC and was taken for a lengthy ride in it by a man who seemed middle-aged and droll, but I suspect now that he was no more than 35years old. What I am more certain of is that he was an incredibly smooth and fastidious driver who regarded 4000rpm as a maximum even for hot-blooded motoring and who thought little of my ideas about open-air motoring. I gained enough from the experience, though, to know that having the speedometer in front of the navigator/passenger was a better idea than having it in front of the driver when it came to working out time and distance calculations.

Despite my lack of wheels, I became involved with the sports car crowd. Not to put too fine a point on it, they were a fairly wild crew. One

certainly was not too drunk to drive if, by holding one hand over an eye, the double vision could be eliminated. There were times, though when even such an expedient could not reduce the number of door handles and steering wheels by the desired 50percent and it was then that I was actually invited to conduct car and owner to a place of recovery. Thus, I got to drive a variety of MGs, but many of them were TCs; some were completely standard, although most were modified, often to the point of lunacy, but that added to it all.

No one ever seemed completely satisfied with TCs. Wheels were reduced in size to take fat 16in tyres (crossply, of course) in the cause of providing better road grip, engines were tuned by the addition of larger SU carburettors, compression ratios were raised by the simple technique of shaving the head, camshafts were reground to outlandish profiles, exhaust systems bent into exotic shapes to make extravagant noises, superchargers were fitted behind bulging bonnet sides. Indeed, some of the cars were even used for hill climbs and races at weekends as well as providing commuter transport for the other five days — that is, when dropped valves and dissolved big-ends didn't preclude the vehicles from taking to the road.

Most of them, of course, never moved closer to the race track than the spectators' car park. But there was another sort of racing, rather more risky, infinitely more daring and far less costly in the long run. That was the race to the spectators' car park, the race from the club meetings to the coffee bars, the race to the social events and the race from the area before police, alerted by those disagreeable people who insist on trying to sleep at 2am, arrived note books and prosecutions at the ready. These races were enormously enjoyable, particularly to one with absolutely no knowledge whatsoever of the disaster potential. When my imagination was finally stirred into the realisation that I was riding not in a car but on borrowed time, my impetuous enjoyment of it all waned somewhat. Nevertheless, it was fun; the search for more modifications and improved driving techniques resulted in a different winner almost every time. The longer hauls out of town proved the worth of the lesser-tuned TCs. Slower they may have been but they looked awfully fast when seen from the side of the road as one pondered how it was that the ground had suddenly become

visible through the spark plug holes of one of the much more powerful but regrettably less reliable versions of the same thing!

There was, frequently, deeper involvement with the faster TCs. There always seemed to be something to do to them to get them back on the road or on the track and usually late at night. There tended, too, to be ugly scenes with neighbours who, having suffered a decibel invasion all day, were distinctly snakey about the whole thing by the early hours of the following morning. But it was respect for the urban complainers that inspired the scheme of tow-starting a just-rebuilt TC around a cemetery perimeter road at 2am. So anxious were we to ascertain the effect of the most recent modifications that we refitted neither bonnet nor exhaust manifolds for the first trial run and, I must admit, the spectre of a crude, bellowing car emitting sheets of blue flame from the exhaust ports as it hurtled around a road given more normally to slower and more sombre processions, casting weird reflections as it went, made us all decide that the wrath of neighbours was preferable to the wrath of wraiths.

That particular TC was half mine, although I seemed to use it somewhat less than my financial stake would suggest. It was British racing green, and well used even when it fell into our hands, but for reasons that were never satisfactorily explained it rode and handled far better than any other TC I drove. A gentle and friendly car, it always behaved itself and accepted alterations both to its engines and its bodywork with a resigned dignity. In the days when it was supercharged with a Roots-type blower it was immensely fast but because it had developed a passion for methanol fuel, which it consumed at an horrendous rate, it had to be detuned for road use. That really meant no more than plopping on a single carburettor manifold from a Y-type saloon. Needless to say, performance was quite hopeless but enough for my co-owner to be caught speeding in it yet again. Instead of the magistrate sighing as he always did and handing *down* a fine, he told the luckless defendent to hand *up* his licence for a few months!

For that very reason, I was conducting my friend and his girlfriend in the TC when we were struck a glancing blow on the left rear wheel by a drunken PR man who had completely overlooked the 'stop' sign. Even though I could see what was going to happen, the single 1.25in SU carburettor just could not provide enough acceleration to get clear. The girl went flying in one direction, me in the other, whilst my friend stayed on board during the ensuing roll-over and return to right-side up. None of us were seriously injured which was really quite miraculous, but the car was a sorry sight. Apart from the dents and scratches, the windscreen pillars had been pushed clean through the scuttle and the hood, which had been erected at the time, was gravel-rashed and forlorn, no longer having a windscreen to grip. Because of financial problems the car was patched up rather than repaired and thereafter ran with a pair of aeroscreens instead of a windscreen. That meant there could not be a hood either, which was more or less okay in summer but left one permanently poised on the brink of pneumonia the rest of the time.

For reasons that are not at all hard to understand, that green TC drew policemen to it like gigolos to the Rich Widow's Ball. Sometimes they could be convinced that all was legitimate but occasionally there was an unseemly scene, such as the time the eager beaver from the motorcycle division insisted that the car was equipped with a bald tyre, having failed to find much else wrong with the vehicle. The TC was actually on the way back from a sprint meeting and to conform with the letter of the competition rules, was carrying a spare wheel — with the oldest, baldest tyre we could find on a 19in wheel as opposed to the 16in wheels at the four corners. I can't quite remember how it all ended, but there was a fierce argument. That particular car, which had passed through many hands before we got it, passed through a lot more afterwards, too. The final owner, so I heard, wrote it off — and didn't have to worry about the loss of a no claim bonus either.

A girlfriend of mine also had a TC. It was immaculate, black with champagne trim, and completely standard. She was some driver and proved to be notably good at hill climbs in the hotter cars she was offered. However, long distance events were not her forte any more than long distance road trips. It was all to do with the TC's ride quality accelerating the kidney process to the point that more time was spent in search of secluded bushes and toilets than actually driving. My girlfriend had a girlfriend who also had a TC, a somewhat older model, and not as spotless. She got it cheap because there was a strange thumping noise from underneath which turned out to be the broken gearbox mount allowing the propeller shaft to touch the tunnel. Another chap and I fixed it for her in an afternoon (if nothing else, TCs were fast to work on; that task involved removing the engine, then the offending mount, taking it away to get it welded and putting the whole lot back together again). That evening was spent in the reckless pursuit of establishing which of the girls had the faster accelerating TC — to the annoyance of the inhabitants of an incredibly snooty suburb, who kept rushing to their gates waving fists and shouting threats. Just as well we were making so much noise we couldn't hear what they were saying. Mind you, noise was about all we were doing. The skinny back wheels would obligingly spin when you dropped the clutch in first gear, but the acceleration was just about on a par with a 1978 Citroen Dyane. No wonder so many of us managed to survive!

The severe space limitations of the TC's cockpit made the car an instant hit with mothers. They could see at a glance that their daughters (and sons, even), were unlikely to be deflowered in such an environent. Yet those slimmer and more athletic than I had a modicum of success in the TC. One of the car's big advantages was that the steering wheel could be pulled from its spline to provide more manoeuvring space. One randy chap

of my acquaintance, having chosen an isolated park for his exploits, was ill-advised enough to put the wheel along with most of both parties' clothing onto the hood in an attempt to increase the amount of interior space available for the athletics meeting in the cockpit. However, during the excitement a passer-by pinched the wheel *and* the clothes. It took quite a long time to find a solution to the problem, but eventually a pair of multigrips from the toolkit provided just enough leverage on the spline to get the girl to her house — by now it was broad daylight — where they were confronted, almost nude, by outraged parents with horsewhips and so on. The thief was never forgiven: the multigrips ruined the spline!

How could any one have less than fond memories of TCs? That's why I bent the ample arm of Chris Harvey, sometimes CAR contributor and author of *The Immortal T Series* (Oxford Illustrated Press Ltd) to find me a bog-standard TC to drive tor an hour or so. Since an immaculate TC now brings as much as a second-hand Ferrari Dino 246, I'm not exactly in the market to indulge my fancy, however desirable it may seem as I recall the glorious exploits of a lost youth. Harvey pointed me in the direction of Alastair Naylor of Naylor Bros, in Shipley, Yorkshire, a firm specialising in remanufacturing MG T-series. They very professionally restore the cars from — and including — the chassis rails upwards, so that their TCs are probably better than brand new when they roll from the works for a 200mile test to ensure that they are, well, new. Closer to specialist-manufacturers than restorers, Naylor Bros have a vast international clientele and a five-year waiting list. My first visit to Naylor Bros was dogged by pelting rain and although it would have been possible to drive a beautifully rebuilt car that was just about to be delivered to a Dutch customer, I decided against it. For this last tango I wanted a fine day so that I could go motoring with the hood down and suffer a little to remind myself that youth and objectivity are total strangers. We agreed another day and so it was northwards once more, this time from Norwich in a Leeds/Bradford bound Air Anglia F27 to meet up with editor Nichols zipping up from London in an Alfa Giulietta 1.6, rather than the Porsche 928 that had carried us both there the first time.

This time Alastair Naylor had found not a pristine TC on which an enthusiast would lavish countless hours of patient polishing, but a genuine unrestored 1946 model. Its owner, Richard Green, bought the TC 12years ago, got involved in other things, namely a Healey 3000, and put the MG aside for six years. Straight from its MoT hours before and a very recent engine rebuild, it sat in a corner of Richard's workshop — the one given over to the maintenance of his fleet of Scania trucks. With its cream body, black wings and well-used appearance, the TC looked honest to the point of innocence. It was almost all there: the hood was erect but the side screens were missing and there was a non-standard four-spoke Brooklands wheel atop the steering column. The door sagged nicely on its hinges as I opened it to climb once more into the passenger's seat. Slight pause while I refreshed my memory as to the technique for making a moderately graceful entrance: bum-in-first seemed to register and it worked. In fact, it was easier than I had expected but Richard and I seemed to fill the cockpit rather more completely than I ever remember two other people doing. Yet we often carried three people in TCs. It must have been immensely uncomfortable. Sometimes a person would actually sit in the luggage compartment behind the seat. They were either very small people or were immune from discomfort. Two friends once found a somewhat slow would-be car thief in the act of nicking their

TC. They were able to jump him from behind, tumble him around a little before tying him securely and tossing him into that little shelf. By the time he had been bounced for 20miles along indifferent country roads to the nearest policeman, he was pleading for incarceration in a nice, comfortable dungeon.

So, when I rode in and drove this amiable Yorkshireman's completely standard TC (none of your telescopic shockers or fat, lower pressure tyres) I had expected that my spine would come up through my skull. As is well known, the TC's beam axles and leaf springs take no account of ride comfort and cars have come on 32years since Richard Green's TC was bolted together at Abingdon from what amounted to a collection of pre-WW2 bits. I felt almost cheated when my toenails didn't suffer from structural fatigue, nor the retinas of my eyes detach themselves. In fact, the ride was perfectly reasonable, far better than I had reckoned possible. Japanese car habitues would probably get into a TC and not notice anything too different about the level of comfort, except that the seats are better.

With the hood up against the cold — still no side curtains — the TC wasn't all that draughty despite those cutaway doors. Their gradual deletion from sports cars really was a retrograde step. Of course, the ergonomics marched to a different tune then, as I was to remind myself as I came to grips with the idea of a long-legged, short-armed (in a manner of speaking) driving position. I have certainly put on weight since I last drove a TC. But the steering wheel was only inches away from my belt buckle this time and my elbows seemed to need quite a lot of room, too. But then, that's what those cutaway doors are all about. Memory easily chops off forgettable items. The MG TC's steering is forgettable. Even making allowance for the fact that Richard Green's car had slightly worn steering, it's still pretty weird by today's standards. There's an awful dead spot in the straight ahead position and since TCs never stay on course for very long you need to tug the big wheel frequently so that progress is... er, well, uncertain until you get the hang of how much of what is required. The brakes were okay way back when.... now they seem quaint and not too effective but that's partly to do with the need to put a fair amount of muscle onto the little pedal. Lovely fly-off handbrake — works like a charm.

Great little gear change too. Tiny knob but a lovely action. No synchromesh on first, but very well selected ratios. Didn't have the heart to press on, for the engine was having its first outing since being renovated. But quite obviously, now, TCs never actually did anything much in standard form. They seemed quick enough at the time, though, and that certainly helped the '40s and '50s youth keep body and soul together in a person-shaped package, I'm both sure and grateful.

Twittering across the moors in that original TC,. slightly battle scarred, slightly tired, slightly behind the times, slightly suspect in steering and brakes, slightly uncomfortable, and more than slightly noisy was enough to make anyone look back on lost youth. Trouble is, you just can't recapture it when the car and the person have both ended up in the same condition.

Meanwhile, the queues are still forming outside Naylor Bros. Customers from all over, cheque books at the ready, are waiting to pay big money for reborn MG TCs (TDs and TFs also, it should be said). I won't speculate as to the reasons; everyone is sure to have a different motivation. If I could afford between £6000 and £8000 for a fully restored TC I'd surely have one — as a little monument to irresponsibility. No, no. Not in the present financial sense. A monument to irresponsibilities past and survived.

Youth at the

helm

It was, of course, the car that started the post-war sports car craze and for a generation you spelled sports car MG TC.
By Geoffrey Bewley.

1946 MG TC.

Youth at the helm

HE MG TC is a marvellous car. It was
planned as a stop-gap, it was 10 years
out of date when it came out and it
led on to the market in small handfuls.
it captured the hearts and minds of
ons and in the end it came to define its
class. A TC was a sports car. Other cars
sports cars to the extent that they
like TCs. For a lot of people this still
true.

young man is first attracted to a girl by
ooks. He finds out that the respects her
acter and admires her mind later on,
he gets closer. The TC's looks do the
trick. It looks light, clean and elegant,
the long straight bonnet and short
it and tail balanced on the sweeping
f the guards. With the tonneau stretch-
ver the passenger seat and the screen
it looks lean and purposeful. Either
it embodies an ideal.

the cockpit, you sit with the wheel up
ur chest, the dash over your lap, and
gearstick, the pedals and your feet a
way down out of sight under the facia.
gear-changing arm's elbow isn't far
your passenger's ribs and your right
rests naturally on the cutaway door.
road ahead is away in the middle
nce, beyond the octagonal radiator cap
e end of the long shiny bonnet hinge.

u start up and enjoy the quick throb of
xhaust for a few seconds. You tap your
on the accelerator and listen to it
ge to a hard, flat, tearing note. It's im-
ble to describe how this makes you feel
e. Then you reach in under the dash,
n the gear and you're pulling away.
cause you're in the open, with the
n down and the wind in your face, you
greater feeling of speed as you go up
gh the gears. The car has a fine, free,
limbed feel under you. The box is very
, with a clean feel as it clicks home and

e TC feels stable under you on the
ers, shivering just a little on its
ry wire wheels."

locks in each time. Once up through it and
it's second nature.

The ride is hard but this gives you an idea
of how you're going. The TC feels stable
under you on the corners, shivering just a
little on its spidery wire wheels. The steering is accurate but on the sensitive side. It
goes where it's told, but you keep having to
tell it. But it does transmit a lot of the feel
of the road up through your hands.

It's really not much of a handful on the
road, because with practice most of it
becomes instinct. You get it right without
worrying about it. You're safe unless you do
something absolutely stupid. But you're not
likely to lose track of what you're doing
while you're enjoying it so much.

"Safety Fast" was the MG motto. MG
history started in 1924, when Cecil Kimber,
the general manager of the Morris Garages
in Oxford, started selling improved examples
of the standard Morris tourers. But MG
sports car history really started in 1928,
when the first MG M-type Midgets appeared
at the 1928 London Motor Show. These
small, neat, handy, cheap two-seaters put
sports motoring within the reach of
thousands of young men who could never
have afforded even a Frazer Nash otherwise.

M-types were tuned and raced, and they
led to a whole string of improved Midgets in
the early '30s. These quickly became more
complex and specialised, and in 1935 Lord
Nuffield (the ennobled William Morris)
ordered a return to first principles, good
straightforward cars with a lot more parts
in common with the family Morris models.

Kimber and his design team chafed at the
restriction, but they planned a new Midget
styled like the handsome and popular J2 of
1932, with a long bonnet, cutaway doors
and a slab tank, with lots of Morris parts
and a new Morris engine. This was the MG
TA.

"By the mid-'30s," one motor historian
says, "the name MG had become as much a
synonym for sports car as Kodak was for
camera or Hoover was for vacuum cleaner."

"It looks light, clean and elegant,
with the long straight bonnet and
short cockpit and tail balanced on
the sweeping line of the guard . . .
it embodies an ideal."

The legion of MG enthusiasts who'd become
used to high-revving overhead camshaft
engines in the previous Midgets objected to
the TA's 1292 cm³ pushrod unit, which
substituted size for refinement. But the TA
was sound and cheap, it won victories in
trials, and soon it was selling well.

The TA's engine was its weak point. It was
a good enough unit for touring, but it wasn't
up to the sort of thing most MG drivers
wanted to put it to and it wouldn't respond
to tuning. In 1939, after 3003 were built,
the TA gave way to the TB with the new
1250 cm³ XPAG engine and the same clean-
cut lines.

It happened that 1939 wasn't a good year
to introduce a new sports car, and only 369
TBs were built before the MG plant was
turned over to war work. Kimber left the
company and the Morris empire in 1941,
and died in a wartime railway accident.

In 1945 it was time to get back to car construction. H.A. Ryder, MG's new managing
director, decided to get cars on the road fast
by putting the TB back into production. The
XPAG engine was retained, the body was
made wider, the suspension was improved,
and it was renamed the TC. By the end of
1945 81 of them had been built.

Ryder's decision to stick with a tried
design was a very wise one. It gave MG an
instant advantage over all other potential
British sports car makers. The TC was handsome and cheap, and that helped too. In four
years of production it renewed and enhanced
the reputation Kimber's cars had won 15
years before.

It had a box section steel chassis with
tubular cross members, with a wheelbase of
2387 mm and a track of 1143 mm. The body
was built of steel on an ash frame, with a
long folding bonnet forward and a tall
61.4-litre fuel tank aft. It came with a red,
black or British racing green finish, with
silver on the 19-inch centre-lock wire wheels.

The XPAG engine was a rugged unit
which would stand an amazing degree of
tuning. The carburettors were twin 1¼-inch
SUs. The compression ratio was raised
slightly, to 7.5:1 from the TB's 7.3:1. The
standard engine developed 40.5 kW (54.4
bhp) at 5200 revs.

The cockpit was cramped at about chest
level, but it suited the short-arm driving
style of the day. There was plenty of leg
room in under the scuttle. There was a flat
shelf behind the seats, above the diff, for
luggage or a third passenger squeezed in
crossways with his knees up. The roof folded
down neatly into the body line. The sidescreens went in a separate space in the back.

It was and is a good car to work on. Stand
the windscreen up and fold back a side of
the bonnet, and everything on that side of
the engine is laid out neatly in front of you.
Fold up both sides and lift it off, and the
whole front of the car between the wings is

Youth at the helm

bared for inspection. You can see at a glance where everything goes. You can change plugs in shirtsleeves without dirtying your cuffs. You can stow tools right where you need them, in a long box built on to the bulkhead.

The TC in these pictures belongs to Sydney car dealer John Thompson. "I've had it about 12 months," he says. "I bought it off Stewart Ratcliff, and he was the one who restored it. It wasn't a good car to begin with, he told me that. It wasn't anything special, a low mileage or anything.

"He had a chassis and an engine and all the various bits and pieces, but there wasn't one complete car. He built it up out of about three cars. He imported some parts, and he built it up out of that and new parts, restored bits and whatever.

"I bought it because I like them, and it was a good one, I'd never seen a better one. And we're in the business, sports cars are our business, and I suppose we've used it in a small way as a promotional thing, just having it around the place."

The used value of TCs in Australia went on dropping until about the middle '60s, when it levelled out. They were technically outdated by then, but they'd always had a hard core of unshakeable admirers and their nostalgic appeal was growing. By 1970 prices for good examples were rising, and they've gone on climbing steeply through the last 10 years. We saw the $10,000 mark for a restored TC reached several years ago.

"I'd sell this one if I had the right price," John Thompson says. "I've found that you're sorry when you sell something, but you can always find another one. I'm not sentimental over cars. If I was sentimental I'd have all the MGs in Australia and no money.

"I'd want twenty thousand for that because I know what it cost to restore. Most of the people who restore them forget the hours they've put into them personally. There was a fellow at the last concours who said his car cost fifteen thousand to restore, and that was without his time. I'm darned if I know. But if you had Ian Cummins restore a car for you, not that he's expensive, he's just charging hourly rates and running round doing all the work, and chasing up bits and pieces, it'd cost you a lot of money.

"There was a TC at the last concours that we sold in 1969 for $1750, and it's still only done about 100,000 miles. I think it might have done about 80,000 miles, actually. It's still in very good original condition, it won an award at the concours for originality. And that fellow wants about $12,000 for it. So they have all appreciated in price."

The value of a lot of good old cars has risen sharply in the last decade, but I can't think of any make that's gone up as fast and

MG's original press handout photograph for the 1948 TC.

as far as the T series MGs after dropping so low. It's not easy to pick the exact reason for this.

Their performance, unmodified, was only just about adequate for their day, and it's very unremarkable now. Their styling was 10 years out of date from the start. They were noisy, hard-riding, and desperately short on passenger comforts. The TC dominated the British and American sports car markets in the '40s through the weight of numbers made, but as a 10-year-old design it was always living on borrowed time.

I think they're treasured today because of their beauty and because of the hold they won over enthusiasts' imaginations. Because Ryder had decided to stick to the basic TB design, MG was able to push straight ahead with mass production. If only 10,002 were built, well, that was still a lot for the time. The MG TC was the only mass-produced British light sports car throughout its production life.

Once again sports car and MG were practically synonymous. The generation that grew up after the war learned to accept spidery wire wheels, flared guards, long straight bonnets, slab tanks and cutaway doors as part of the definition. By the time streamlined modern sports cars reached the market, the TC had already shaped men's minds for good and ever.

It probably wouldn't have worked out so easily if the TC hadn't been so beautiful to start with. But it was the last and finest example of classic sports car styling, as the China clippers were of sailing ship design. Everything came together perfectly just before the age died.

They were the sort of car driven by cheerful, clean-cut Battle of Britain pilots in the movies. They epitomised hardihood and

"It was the last and finest example of classic sports car styling, as the China clippers were of sailing ship design. Everything came together perfectly just before the age died."

sportsmanship. Driving one fast you could feel like one of your boyhood heroes, Segrave or Moss or Biggles, as you crouched in the open cockpit with the sun on you and the exhaust note whipped away by the wind.

They tended to be long-lived, passing from owner to owner down the years. It's only now that a generation is growing up that won't be able to cut its motoring teeth on them. Today they're nearly all restored and stashed away safe in garages as weekend outing cars. Their numbers will increase slowly as the last few goers are put together out of rusty chassis and old spares, but there'll never be enough of them to go around.

"You sell one and you think, I'll never get another one like that," John Thompson says, "and sometimes that has happened. I've had nice cars and I've sold them and I've thought, I'll never replace that. And a couple I haven't, but there's others you can. There are a couple of TCs around as good as that one, that I've seen. Two very good red ones were at the last concours.

"There's no hope for British sports cars today. People want British sports cars, but somebody's doing something wrong. The poor old Poms are going about things the wrong way, I think. They had it and they've given it away."

It's probably not surprising nobody's mass-produced a car as attractive as the TC since, but it's a bit sad that nobody today seems to be trying. It would be nice if it was a mark people kept on aiming for, not just a masterpiece of a lost age.

MG **Naylor Brothers** **MG**

FAMOUS FOR QUALITY
RENOWNED FOR VALUE
WORLDWIDE IN SERVICE

BL Heritage
approved supplier
and restorer
MG T-type Midgets.

MOTOR
AGENTS ASSOCIATION

Over 1,300 superb
quality T-type spares
are listed in our
illustrated catalogue
(£1.25)

The MG T-type
specialists with a
worldwide
reputation for
Concours rebuilds
and spares service,
now offer the same
high standards of
mechanical and
bodywork rebuilds
to owners of the modern MG Midget,
MGB, MGC and V8.

Due to the recent expansion of our
workshops, we are also prepared to
accept any other classic marque for
Concours standard paintwork.

Naylor Brothers, Airedale Garage, Hollins Hill, Shipley,
West Yorkshire BD17 7QN, England.
Telephone: (0274) 594071 Stores, 585161 Office

Access, Barclaycard, American Express Welcome.

BL Heritage
Approved Supplier

BROOKLANDS TECHNICAL BOOKS

Brooklands Technical Books has been formed to supply owners, restorers and professional repairers with official factory literature.

We have at present over 250,000 handbooks, spare parts catalogues, workshop manuals, sales brochures and other original material in stock.

The following is available on the MG TC:

MG TC INSTRUCTION MANUAL
Covering maintenance, lubrication,
chassis, brakes, steering, engine,
carburation, gearbox electrical
systems and coachwork.
  104 Large Pages

Instruction
Manual
for the
MG
Midget
(Series "TC")

Our inventory is changing daily, please write for our latest list or telephone for specific titles.

BROOKLANDS
TECHNICAL
BOOKS

Brooklands Books Distribution Ltd.
Holmerise, Seven Hills Road,
Cobham, Surrey KT11 1ES
Tel: Cobham (09236) 5051

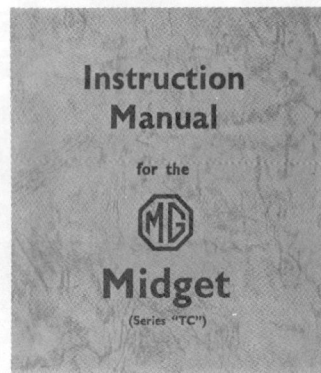

MG
TC-TD-TF PARTS

Join the thousands of satisfied T-Series owners around the Globe who rely on us for quality spares to restore and maintain their cars. Our 65-page, fully-illustrated Catalog / Parts Manual is the most comprehensive volume in the market. Send $5.00 in U.S. funds and select your needs from the 3,100 different items made available by the world's largest T-Series Parts Specialist. Our *only* business is making parts fit to a T!

The company run by enthusiasts – for enthusiasts.

Abingdon Spares, Ltd.

**P.O. BOX 37(B), SOUTH ST.
WALPOLE, NH 03608 U.S.A.
TEL. 603-756-4768**

WE WELCOME MasterCard VISA

FAST FAST SERVICE MAIL ORDER
(CATAOLG AVAILABLE UPON REQUEST)
800-631-8990 8991 (USA)

VISA   MasterCard

MG – TC – TD – TF
MGA
MGB

MG – TC – TD – TF
MGA
MGB

PARTS MAIL ORDER
SALES SERVICE

FAST FAST SERVICE

THIS YEAR, BUY ALL OF YOUR MG'S NEEDS FROM THE SOURCE! WE CARRY A FULL LINE OF ASH WOOD COACHWORK, BODY PARTS, MECHANICAL PARTS, INTERIOR KITS, TOPS & TONNEAUS. WE HAVE A REBUILDING PROGRAM FOR SHOCKS, GAUGES, AND JUST ABOUT ANY ITEM ON YOUR MG! SO THIS YEAR, COME TO THE SOURCE!

"Don't substitute cheap Hong Kong specials for original British parts".

Buy British   Buy from the source

Buy from

M & G Vintage Auto Co.

**154 CHESTNUT STREET   RIDGEWOOD, N.J. 07450 USA
THE FIRST BRITISH LEYLAND HERITAGE RESTORER IN THE USA**

(CATALOG AVAILABLE UPON REQUEST)

FAST FAST SERVICE

MG – TC – TD – TF
MGA
MGB

MG – TC – TD – TF
MGA
MGB

MG TC 1945-1949

Some 35 stories from Britain the US and Australia are brought together to form a profile of the first post-war MG Midget, the TC. Included are road tests, used car tests, driving impressions, plus articles on touring, record breaking, racing, restoration, rebuilding, and repair in general. A 1000 mile test is undertaken which ends at the factory at Abingdon. More recent articles written during the 70s and 80s cover the TCs current position and comment on how it fits into the classic car scene.
100 Large Pages.

MG TD 1949-1953

The development of the MG TD is trace through 34 articles drawn from the leading journals of Australia, the US and Britain. Included are road tests on the Series I and II, a comparison test v. the TC, 2000 and 35,000 mile reports, together with stories on tuning, touring, rebuilding, racing, and record breaking. A used car test is also reprinted plus a fascinating report on a journey from Bombay to London.
100 Large Pages.

MG TF 1953-1955

Reports from Britain, the US, Ireland and Australia lead us through the production life of the TF Midget. Articles cover road tests of the 1250, 1500 and supercharged cars, comparison tests against the TC and TD models, new model announcements, plus stories on tuning, racing, restoration, history and a visit to Abingdon. Also included is an 8 page data sheet outlining all servicing requirements of the TF.
100 Large Pages.

These soft-bound volumes in the 'Brooklands Books' series consist of reprints of original road test reports and other articles that appeared in leading motoring journals during the periods concerned. Fully illustrated with photographs and cut-away drawings, the articles contain road impressions, performance figures, specifications, etc. None of the articles appears in more than one book. Sources include Autocar, Autosport, Car, Car & Driver, Cars & Car Conversions, Motor, Motor Racing, Modern Motor, Road Test, Road & Track and Wheels. Fascinating to read, the books are also invaluable as sources of historical reference and as practical aids to enthusiasts who wish to restore their cars to original condition.

From specialist booksellers or, in case of difficulty, direct from the distributors:
BROOKLANDS BOOK DISTRIBUTION, 'HOLMERISE', SEVEN HILLS ROAD,
COBHAM, SURREY KT11 1ES, ENGLAND. Telephone: Cobham (09326) 5051
MOTORBOOKS INTERNATIONAL, OSCEOLA, WISCONSIN 54020, USA.
Telephone: 715 294 3345 & 800 826 6600

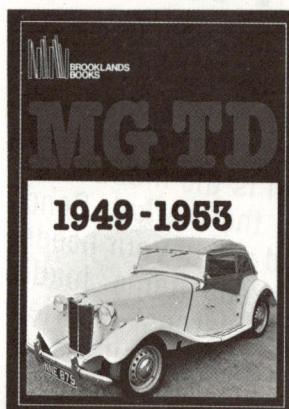

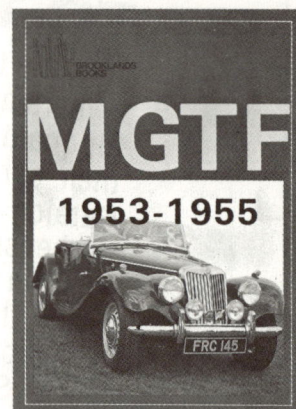